Pregnancy, Risk and Biopolitics

Historically, western cultures have considered that human status was conferred at the end of childbirth. However, in the contemporary medicine, law and politics of the global north, the living subject is often claimed to pre-exist birth. In *Pregnancy, Risk and Biopolitics* Weir argues that the displacement of birth as the threshold of the living subject occurred in the 1950s with the novel concept of "perinatal mortality" referring to death of either the fetus or the newborn just prior to, during and after birth.

Pregnancy, Risk and Biopolitics calls attention to the significance of population politics, especially the reduction of infant mortality, for the unsettling of birth as the entry to human status. Weir traces the introduction of a new perinatal threshold into child welfare and tort law through expert testimony on fetal risk, sketching the clash at law between the birth and perinatal thresholds of the living subject.

This novel book makes original empirical and theoretical contributions to the history of the present (Foucauldian research), feminism, and social studies of risk. Weir conceptualizes a new focus for the history of the present: the threshold of the living subject – the historically and culturally variable processes through which the boundary of human status is established at the points of entry and exit into collective existence. She argues that the risk techniques are heterogeneous, subjects of contention, and plural in their political effects rather than singular, neoliberal, and uniformly supported by expertise.

This book is a well researched and accessible study of biopolitics which is of interest to students and researchers in anthropology, health studies, history, legal studies, science studies, sociology, and women's studies.

Lorna Weir is an associate professor in the Department of Sociology at the York Centre for Health Studies at York University, Ontario, Canada, and is a member of Health Care, Technology and Place, Canadian Institutes for Health Research (University of Toronto).

Transformations: Thinking Through Feminism

Edited by

Maureen McNeil, *Institute of Women's Studies, Lancaster University*
Lynne Pearce, *Department of English, Lancaster University*

Books in the series include:

Transformations
Thinking through feminism
Edited by Sara Ahmed, Jane Kilby, Celia Lury, Maureen McNeil and Beverley Skeggs

Thinking Through the Skin
Edited by Sara Ahmed and Jackie Stacey

Strange Encounters
Embodied others in post-coloniality
Sara Ahmed

Feminism and Autobiography
Texts, theories, methods
Edited by Tess Cosslett, Celia Lury and Penny Summerfield

Advertising and Consumer Citizenship
Gender, images and rights
Anne M. Cronin

Mothering the Self
Mothers, daughters, subjects
Stephanie Lawler

When Women Kill
Questions of agency and subjectivity
Belinda Morrissey

Class, Self, Culture
Beverley Skeggs

Haunted Nations
The colonial dimensions of multiculturalisms
Sneja Gunew

The Rhetorics of Feminism
Readings in contemporary cultural theory and the popular press
Lynne Pearce

Women and the Irish Diaspora
Breda Gray

Jacques Lacan and Feminist Epistemology
Kirsten Campbell

Judging the Image
Art, value, law
Alison Young

Pregnancy, Risk and Biopolitics
On the threshold of the living subject
Lorna Weir

Pregnancy, Risk and Biopolitics

On the threshold of the living subject

Lorna Weir

Routledge
Taylor & Francis Group
LONDON AND NEW YORK

First published 2006
by Routledge
2 Park Square, Milton Park, Abingdon, Oxon OX14 4RN

Simultaneously published in the USA and Canada
by Routledge
270 Madison Ave, New York, NY 10016

Routledge is an imprint of the Taylor & Francis Group, an informa business

Typeset in Times by
Rosemount Typing Services, Auldgirth, Dumfriesshire

Printed and bound in Great Britain by
The Cromwell Press, Trowbridge, Wiltshire

British Library Cataloguing in Publication Data
A catalogue record for this book is available from British Library

Library of Congress Cataloging in Publication Data
Weir, Lorna, 1952-
Pregnancy, risk, and biopolitics : on the threshold of the living subject / Lorna Weir.
p. cm.
Includes bibliographical references and index.
ISBN10: 0–415–39257–8 (paper back) – ISBN10: 0–415–39258–6 (hard back)
1. Prenatal care–Social aspects. 2. Fetus. 3. Fetal–Legal status, laws, etc.
4. Biopolitics. I. Title.
RG526.W45 2006
362.198'2–dc22

2006004013

ISBN10: 0–415–39258–6 (hbk)
ISBN10: 0–415–39257–8 (pbk)

ISBN13: 978–0–415–39258–7 (hbk)
ISBN13: 978–0–415–39257–0 (pbk)

Contents

Figures

Acknowledgements

A number of friends and colleagues provided steadfast encouragement while I was at work on this book. Carolyn Strange lent a willing ear on many occasions, patiently instructing me in the importance of reconciling scholarship and publicity, titles and electronic database searches. Bruce Curtis radiated calm confidence at crucial junctures. I am hoping to repay Eric Mykhalovskiy for his long patience and comments on the manuscript over the course of our current joint research on global public health surveillance. Pat Elliot read and commented on the first chapter, assuring me it made sense. Fiona Miller and Tom Loebel generously read part of the manuscript. Gottfried Paasche improved translations, and Barbara Godard supplied emergency advice in literary theory.

Paul Datta's critical comments tried to save me from empiricism. The manuscript benefited from his critique, and I continue to contemplate the status of theory in theoretico-empirical work. The preliminary theoretical observations on sovereignty found here are indebted to recent writing with my colleague Brian Singer. Barbara Duden gave the penultimate draft a welcoming reception. Her remarks on the manuscript strengthened the final version, making the process of revising more intellectually interesting.

Many thanks to the feminist legal scholars Shelley Gavigan, Joan Gilmour, Mary Lou McPhedran, Pam Shime and particularly Karen Schucher for their help at various points, though they remain entirely blameless for the resulting treatment of law. The reference librarians at the York University Law Library were extremely helpful with requests for locating sources.

Research on this book began when I was affiliated with the Centre of Criminology, University of Toronto, while on sabbatical. I thank Clifford Shearing and Mariana Valverde for facilitating my initial research while at the Centre, and especially Mariana for her observation that I needed to consider the question of legal evidence more closely.

The contributions of graduate students to the life of the university are many. They are sometimes the people who have the most ongoing engagement in faculty research projects. Lachlan Story provided research assistance for the midwifery and perinatal chapters, and was the only one in my life for an awfully long while who had any interest in the recondite concept of perinatal mortality. It is a pleasure to thank him for his convivial research support. Patti Phillips helped track down citations on short notice. I am grateful for her astute editorial observations, which especially helped the chapter on midwifery. Bill Leeming provided highly skilled research on case law. I am fortunate to hold an appointment in a department (Sociology, York University) that has a strong graduate programme with singular, gifted students.

I thank the midwives and physicians who consented to be interviewed for the gift of their time. My writing here on midwifery was much improved by the remarks of Holliday Tyson to whom I am in many ways grateful for rendering my understanding of midwifery more careful and complex. While I write as a social scientist, I hope my work here will be of interest to health professionals, particularly midwives. Heidi McDonell listened and listened as I talked and talked. Gustava Zhuravskaya kindly tracked down sources in Vienna. Marg Phillips did fine work on the pilot interviews for what became the chapter on midwifery, health and risk.

I have been gifted with a fine editorial process at Routledge. Many thanks to Maureen McNeil, Lynn Pearce and other members of the editorial board for the *Transformations: Thinking Through Feminism* series for their enthusiasm in accepting the manuscript. I much appreciate the intellectual generosity of the anonymous reviewers who lent their minds and time to the peer review process.

And lastly I am glad to thank my sister Lorraine Weir for her comments, intellectual and personal. Many times in our conversations she burst into gales of laughter at my mention of antenatal records and fetal apprehensions. It sometimes takes a *Finnegans Wake* scholar and literary theorist to appreciate fully the paradoxes found in the biopolitical discourse of our present.

Lorna Weir
Toronto

1 On the threshold of the living subject

The threshold of the living subject constitutes the zone of transition into and out of human bodily substance. Unlike the threshold of exit, the threshold of entry occurs in and through another body. Women in pregnancy bear the between, the entrance across which the unborn must pass in order to be distinguished from those who carry them. Neither a pure moment in time nor a single point in space, the threshold of the living subject has duration and extent. When and where that between begins and ends, what status the bodily substance at that threshold might have, is an entirely social and cultural matter.

Thresholds mark the transition from inside to outside, the imperceptible to the perceptible, the non-reactive to the reactive. "The threshold (*die Schwelle*) bears the between" wrote Heidegger (1971: 201) in his essay, "Language".[1] Being the "settlement of the between," the threshold "sustains the middle in which the two, the outside and the inside, penetrate each other" (Heidegger 1971: 201).[2] The between, however, may be difficult to conceptualize for cultural reasons. As Irigaray (1985) argued in her commentary on Plato's myth of the cave, Western philosophy – and other truth regimes – have been incurious about zones and techniques of transition due to their persistent habit of thinking in dualistic categories such as inside and outside, light and dark, truth and illusion. At the birth of Western philosophy, Plato ignored the entrance to the cave.

Without a threshold, a building could not be entered or left; there would be no passage between inside and outside. The most ancient meaning of "threshold" in English pertains to dwellings: "the piece of timber or stone which lies below the bottom of a door, and has to be crossed in entering a house" (*Oxford English Dictionary*, second edition 1989). The first entries for "*sueil*" and "*Schwelle*" in the French and German historical dictionaries, *Littré* and *Duden*, also refer to the wood or stone at the base of a doorway passed over on arrival. Over the past thousand years the meaning of threshold extended to figure the border separating regions or fields, the beginning of an action, and the lower limit of perceptivity or reactivity. A threshold makes possible a relation between heterogenous places, practices and perceptions.

Until the mid-twentieth century in Western Europe and the territories of the European diaspora, the threshold of the living subject had been stabilized in common culture and expertise as definitively crossed only at the end of the birth process. Although its arrival would often be expected during pregnancy, the living subject entered social recognition only at the end of labour, classified then as a child and kin member. However, knowing a woman to be pregnant was fraught with interpretive uncertainties and ambiguities, both from the perspective of the woman herself and from those having expertise in attending births. Women looked to the experience of a new kind of movement in their bodies to know and announce that they were "quick with child". Yet childbirth could occur without the prior experience of quickening, and quickening-like feelings took place without leading to childbirth. Sometimes the sensation of internal movements might cease, leading to further uncertainty; until the mid-twentieth century, physicians considered definitive diagnosis that the unborn was no longer alive difficult to make. Despite the intrinsic ambiguities of the birth threshold with respect to its beginning, it did have a clear end: the birth of a baby, who, if alive, was culturally recognized as having human status, simultaneously a person and an individual.

During the first two decades of the twentieth century, the birth threshold of the living subject came unsettled under the impact of concerted attempts to lower infant mortality and morbidity in Western Europe and North America. The optimizing of health in the first year of life led to the medical invention of a novel *perinatal* threshold, at variance with the previous birth threshold: a problematization of the threshold of the living subject as a site of questions, difficulties, responses. The birth and perinatal thresholds entered into conflict, opening to differing regimes for the governance of the body during pregnancy and labour.

The beginnings of the perinatal threshold took place in European and North American medicine during the period 1920–1950. As infant mortality rates fell during the first two decades of the twentieth century, analysis showed that infant deaths remained high close to birth. Reasoning that deaths prior to, during and shortly after birth had similar causation, and that the bodies of the fetus late in pregnancy and the newborn were fundamentally alike, physicians proposed a new measure and target of intervention: fetal and neonatal mortality, or, as it came to be called during the 1950s, "perinatal" mortality, that is, mortality around birth. The perinatal threshold lightened the significance of birth, consolidating the time late in pregnancy, during labour and just after birth as one continuous interval based on commonalities of bodily substance. The time and space of the perinatal ran across birth.

A threshold stabilizes and sustains a relation of inside and outside: "What goes out and goes in, in the between, is joined by the between's dependability ... The settling of the between needs something that can endure" (Heidegger 1971: 201). Any change in the threshold, the between, alters the relation of inside and out. In the unsettling of the birth threshold by the perinatal threshold, the relation between the inside and outside of pregnancy was put in question. The threshold of the living subject had come to lack dependability.

The perinatal threshold folded a new division of time and bodily substance into the maternal body during pregnancy and birth. The concept of the perinatal distinguished continuities of time and bodily substance for the living subject before, during, and after birth from the time and bodily substance of the pregnant woman. In so doing, it consolidated the existence of the living subject prior to and during birth, providing a rationale for its care: the conservation of fetal life so as to optimize infant health. Where previously the birth threshold only definitively concluded at the end of the birth process with the separation of mother and child, the perinatal threshold distinguished mother from the unborn *during* pregnancy and birth. When the birth threshold became unfixed, the relation between inside and outside at the threshold of the living subject was shaken, affecting the pregnant woman, the living subject, the relation between these two subjects, and the forms of expertise acting on them.

Both the birth threshold and the perinatal threshold have contained interpretive ambiguities. From the perspective of the birth threshold, the status of the unborn between a woman's declaration of quickening and birth was clouded over with uncertainty, a subject about which midwives and physicians knew little, but one that crystallized into a living subject at birth. The concept of the perinatal occupied this ambiguous interval, constituting it as medically actionable: the time of a living subject in some senses distinct from the pregnant woman. The singularity of the perinatal threshold in turn rendered unclear the relation of a woman during pregnancy and the birth process to the living subject she carried. So too, the absent "perinatal subject"[3] during pregnancy and birth had an uncertain relation to the social and legal categories of person and individual. The temporal and spatial distinctions associated with the medical concept of the perinatal, the unsettling of inside and outside at the threshold of the living subject, were thus permeated with gendered power effects.

Thresholds have techniques facilitating entry and exit, techniques which organize the relations between inside and outside, before and after.[4] At the threshold of the living subject, risk techniques were attached to pregnancy and childbirth during the 1950s with the intent of reducing perinatal mortality and morbidity. Standardized, population-based, routine risk assessment in clinical practice came to saturate pregnancy in succeeding decades, promising an ever receding utopia of health. I will show how risk-based prenatal care was offered as a solution to decreasing the perinatal mortality rate by acting on the period of time prior to the birth of the living subject. Perinatal risk factors were folded into the previous method of prenatal care. The calculative, preventive orientation of risk had its basis in epidemiology rather than the actuarial calculations of insurance; risk techniques have been analytically heterogeneous, although uniformly about security whatever their siting. In clinical practice, risk-based prenatal care bound together categories of epidemiological risk with diagnostic information, test results and patient histories; the result was to make standardized prenatal risk assessment into a higgledy-piggledy concatenation of epidemiological and clinical reasoning as risk came to invade the space of patient management, treated as equivalent to any clinical intervention. What goes by the name of risk in prenatal care has thus not strictly speaking been confined to risk judgements. The schema "prenatal risk assessment", in all its analytic heterogeneity, has acted as a transmitter of security for the unborn configured as fetus, standardizing and concerting care of *fetal* health in clinical practice.

The perinatal threshold was consolidated in health care through risk techniques, later attaching to the legal regime through expert medical evidence. In the last two decades of the twentieth century Canadian courts, paralleling other Euroamerican legal regimes, considered displacing the birth threshold of the legal person (other than corporations) with a perinatal threshold in damage claims for prenatal injuries and child welfare litigation. In these actions medical evidence of risks to the health of the unborn was necessary for proof of fact, which establishes whether an event has occurred or a particular entity exists. Damage claims for prenatal injuries were made on behalf of a child born alive with health problems caused by an accident during pregnancy. Recoveries for prenatal injuries crucially depended on medical evidence linking a cause before birth with injury to a child's health after birth. Prior to the 1990s, litigation involved third parties, that is, people who had injured a woman during pregnancy, but in the late 1990s, claims were brought against women themselves for conduct during pregnancy that was possibly linked to fetal health risks. A series of actions launched by child welfare authorities during the 1980s and 1990s similarly used medical evidence of fetal health risks from maternal conduct in an attempt to extend their child protection mandate to the unborn. Child welfare authorities sought to have women ordered into treatment or if necessary detained for the sake of fetal and child health. At issue in these cases was the relation between the child and the unborn configured as a fetus across the perinatal threshold.

Perinatal mortality is empirically associated with poverty, and thus the perinatal mortality rate has fallen on fundamental social cleavages. The series of child welfare cases in the Canadian jurisdiction primarily involved Aboriginal women and coincided with the efforts of health governance during the 1980s and 1990s to reduce the higher comparative perinatal mortality and morbidity rates among Aboriginal peoples. Within health governance, perinatal risk reduction took the form of removing Aboriginal women late in pregnancy from Northern Canadian reserves in order to give birth in Southern hospitals. Perceived as a pool of perinatal risk in health governance, Aboriginal peoples have a specific and harsh historical experience with the perinatal threshold.

As the perinatal threshold spread through risk techniques in the health and legal regimes, it met with a critical reception on the part of expertise. Within the health regime, risk-based prenatal care was criticized for increasing intervention rates with no basis in evidence. Intervention in turn increased iatrogenic effects: the negative health consequences caused by diagnosis and treatment. Midwives and a minority of physicians defended pregnancy as a state of health, priorizing clinical judgement over risk judgement, trying to fashion an alternative basis for pregnancy care separate from risk-based prenatal care: health beyond risk. In litigation pertaining to child welfare and damage claims in prenatal injuries, courts at the appellate level denied the applications that would have made pregnant women accountable for fetal health risks. Judicial law reaffirmed the birth threshold, thereby externalizing the perinatal threshold to the health complex. Unfixing the threshold of the living subject did not result in a rout for the perinatal threshold.

My work here falls within the history of the present – analysis that is broadly foucauldian in inspiration – a school that is deeply curious about the biopolitics of modernity, specifically the forms of reasoning, techniques and power used to optimize the health of populations. The present work extends this inquiry in the direction of examining the new intensities of care coursing around the living subject in the twentieth century at the threshold of its entry into population. Problematizing the threshold of the living subject was an event in biopolitics that shook the grid of intelligibility of the living subject at the threshold of its entry into population, together with the legal address of pregnancy and the liberal governance of women. I undertake a genealogy of a new medical concept, the *perinatal*, and the formulation of a new kind of mortality, *perinatal mortality*, the reduction of which was taken as a goal by national, regional, and local health care systems. Techniques for the reduction of perinatal mortality included a risk-based system of prenatal care intended to promote the security of the fetus. This vast and detailed implanting of a perinatal threshold troubled the previous cultural threshold of the living subject: birth. From the perspective of the birth threshold, the perinatal threshold was a place of paradox: mortality statistics not requiring live birth, antenatal records, and women incarcerated to reduce health risks to those not born.

Risk techniques are analytically heterogeneous, socially contested, and politically variable in effects. My account underscores the heterogeneity of risk techniques and their effects, a writing strategy consonant with current scholarship by historians of the present that seeks to understand risk governance as more than singular and actuarial. I argue that expertise has not simply proliferated risk governance, it has also sought to contain and reject the spread of risk. Exactly how risk governance has been limited or rejected deserves treatment, partly as a corrective to the social scientific writings that have assumed risk roams unchecked over the plains of the present. Nor does risk have any intrinsic belonging to liberal/neoliberal governance; risk is a technology of security with a rich history of both liberal and authoritarian service. Chapter 5 of this book examines an instance of risk taking an authoritarian turn in a series of legal cases brought by child welfare authorities attempting to throw women into detention and coerced medical treatment in order to lower risks to *fetal* health.

The problematization of the threshold of the living subject mobilized four differing forms of power: the security (of population), discipline (of individuals), the sovereign power of law, and governance. These comprise what I will call the power field of biopolitics, an unstable conjunction of interacting powers. Biopolitics may operate within law or where law has been suspended; if within law, biopolitics may be liberal or authoritarian. When the perinatal threshold entered law, the disciplinary confinement and mandatory medical treatment of women for the sake of reducing fetal health risks from their conduct, that is, for the security of the unborn configured as a fetus, was considered and eventually rejected in the Canadian jurisdiction, on the grounds that it threatened the intelligibility of tort law. In the United States of America there has as yet been no determinative reading at the level of the US Supreme Court. The potentialities of the perinatal threshold were given in the power field of biopolitics and have many possible ways of being realized.

My inquiry into biopolitics at the threshold of the living subject arises in the context of longstanding feminist concerns with the "public fetus" (Duden 1993: 50–55) in contemporary gender politics, specifically narratives of maternal–fetal conflict that position women and the unborn as equal persons with competing interests. Contemporary feminism has been haunted by the figure of maternal–fetal conflict and the attempted invasions of women's liberty rights with which that figure had been associated at law. I share these concerns, albeit in the recondite analytic conventions found in the history of the present, but have wondered how the fetus came to be invested with such significance. One might respond to this question with an account of social movements, Christian religious fundamentalism, and discourses of state, but such approaches presuppose a preexisting salience of the fetus within health regimes that these social forces receive and that provides the basis for their claims-making. The extent to which pregnancy has become implicated in risk governance has likewise received feminist commentary, but how this came to pass has had no sustained treatment.

The framing of the present project within the biopolitics of population provides an account of how the unborn configured as fetus came to be publicized and the conditions for maternal–fetal division established. The perinatal threshold marked an historical break from birth as the entry into human status, a break based on a conceptual innovation: the constitution of a temporal interval encompassing newborn child and intrauterine fetus in a common bodily substance. This historical break with the birth threshold was thus not primarily centred in visual imagery pertaining to the fetus, significant though this has been, but in an epistemological rupture that later came to serve a governmental end. When the analysis of infant mortality statistics and the findings of pathological anatomy constituted early infant deaths and stillbirths as preventable, the project of increasing popular health in the first days of life unsettled the birth threshold. By the mid-1950s a programme of optimizing population health over the perinatal interval had been sketched: systematic prenatal risk assessment for the reduction of perinatal mortality rates. At the point that the concept of the perinatal was joined to a technology of risk, the biopolitics of population transcended birth and came to include the unborn. Pregnancy became a time for routinely conserving and optimizing population at and prior to birth. The perinatal threshold problematized birth as the beginning point of the living subject, radiating uncertainty into both common culture and expertise, providing the condition of existence for legal conflict posed in terms of conserving fetal health vs. governing women through freedom.

The story told here begins in Chapter 2 with a genealogy of the perinatal interval: its invention as a concept and the creation of standardized prenatal risk assessment as a technique for reducing the perinatal mortality rate. The discussion then moves in Chapter 3 to an exploration of the professional and ethical conflicts experienced by midwives in risk-based prenatal care. The objections of midwives to risk-based prenatal care display its limitations from the perspective of a professional ethos that priorizes clinical reasoning over risk techniques. In the second part of this book, I examine the transmission of the perinatal threshold into law. Chapter 4 is the first of the two studies dealing with law. It explores the reality effects of perinatal risk evidence on the legal fiction of the child *en ventre sa mère* (the child in the womb/belly of its mother), a concept dating to Ancient Roman estate law that enabled a child born alive after a parental (primarily paternal) death to be treated *as if* s/he had been in existence at the time of that parent's death. This fiction permitted a posthumously born child to become a beneficiary of his/her parent's estate, despite there being no mention of the child in that parent's will. The reality effects of prenatal risk would have stripped the legal fiction, the child *en ventre sa mère*, of its rhetoricity, equating it with a fetus to which women owe a duty of care during pregnancy. In Chapter 5, the second consideration of the perinatal threshold at law, I investigate attempts by Canadian child welfare authorities to extend their legal mandate to include the unborn in order to protect its health and reduce the numbers of children in permanent custody. Child welfare authorities sought to control the conduct of pregnant women in labour, conduct that they argued placed the unborn/fetus at risk. They launched legal actions against women who were for the most part Aboriginal, a population vulnerable to the authoritarian stripping of liberal freedoms given the persistent uses of confinement against Aboriginal peoples in the colonial history of the Americas. This book offers a tale of expertise both inciting and objecting to risk together with authoritarian health governance considered and rejected: biopolitics with complexity.

Biopolitics

"Biopolitics" gestures to a large terrain. After the writings of Agamben (1998) and Foucault (1997a and 1978: 135–159), it has become conventional to understand the conceptual relation between life and politics as having undergone a fundamental change between the ancient and modern periods. The Ancient Greeks had two concepts for what in English is known as "life": *zoē* (bare life), the life common to humans, animals and gods, and *bios*, specifically human ways of life pertaining to individuals or groups. The Ancient Greeks and Romans held that politics was the elaboration of a cultural form of life that was based on, but surpassed, bare life; the moderns made the security of bare life a goal of politics, in so doing giving politics responsibility for population and health.[5] As the medical historian, George Rosen (1958: 17), observed prior to Agamben and Foucault, "[t]he protection and promotion of the health and welfare of its citizens is considered to be one of the most important functions of the modern state." Health in the modern period became a constituent and an objective of politics, a responsibility of governance to improve for reasons political, economic and ethical.

Research on the terrain of biopolitics continues the problematic of health and truth begun in Foucault's *Naissance de la clinique* (1963) and *The History of Sexuality*, Vol. 1 (1978) [1976]. Historians of the present are currently fashioning a research trajectory in biopolitics that is strongly concentrated in the field of genetics (Hedgecoe 2004; Rabinow 2002; Rabinow 1996b; Rabinow and Rose 2003; Rabinow and Dan-Cohen 2005; Rose 2001), with a just emerging literature on bioethics (Memmi 1996; Rose and Novas 2004). What "biopolitics" might be taken to mean is, however, currently uncertain, a situation to which Paul Rabinow and Nikolas Rose (2003) have responded by suggesting an analytic distinction be made between "biopower" and "biopolitics".

Rabinow and Rose formulate biopower as a terrain of rationalized activities that take as their object vital events pertaining to humans individually and collectively (Rabinow and Rose 2003: 2–3). "Vital" carries the sense of human bodily substance in its coming into being, maturing and passing away. Rabinow and Rose characterize biopower as composed of three elements: (1) truthful discourses about vital events made by designated authorities; (2) strategies having the objective of implanting life and health; (3) modes of subjectification through which "individuals can be brought to work on themselves" for the sake of individual or collective health (ibid.: 3–4). "Biopolitics" in Rabinow and Rose's analysis is reserved for the second level of biopower: strategies and contestations over interventions in human vital events.

Rabinow and Rose's treatment of biopower and biopolitics, a much-needed intervention to clarify terminology in a rapidly expanding field, revises Foucault's earlier work in *The History of Sexuality,* Vol. 1 (Foucault 1978 [1976]) and *"Society Must Be Defended": Lectures at the Collège de France, 1975–1976* (Foucault 2003 [1997b]). While Rabinow and Rose constitute biopower as a single form of power, and insert biopolitics as one of its elements, Foucault defines biopolitics/biopower more narrowly and power over life more broadly than they. In both these works, Foucault was concerned to develop a distinction between biopolitics/biopower and discipline while conceiving both as vital. Discipline and biopolitics are, he argues, different power apparatuses, with discipline aimed at the bodies of individuals and biopolitics at the vital events of human populations.

In *The History of Sexuality*, Vol. 1, Foucault differentiates "anatomo-politics" from "biopower" (Foucault 1978: 140). Anatomo-politics is conceptualized as a power for training individual bodies through continuous observation and force in order to optimize their utility and docility. Contrastive to anatomo-politics, biopower targets populations rather than individuals, extending "the order of knowledge and power" into longevity and mortality, public health and migration: "phenomena peculiar to the life of the human species" (Foucault 1978: 141–142). The distinction between power that takes as its object the individual body and power that aims to act on the vital events of populations by intervening in the birth rate, life expectancy and so forth, is again forcefully emphasized in *"Society Must Be Defended": Lectures at the Collège de France, 1975–1976* in the lecture of 17 March 1976. Foucault there repeatedly emphasizes that "biopolitics" fashions its terrain as "aleatory events that occur within a population that exists over a period of time" (Foucault 2003: 246), random occurrences that, conceptualized through abstractions such as "mortality", may be modified through "regulatory mechanisms". Biopolitical mechanisms intervene in rates, equilibria, and series, acting to "maximize and extract forces", but "[u]nlike disciplines, they no longer train individuals by working at the level of the body itself" (ibid.). In an important observation, Foucault asserts that biopolitical mechanisms (which "apply to the population as such" (Foucault 2003: 252)) and disciplinary mechanisms connect through the norm, acting together to constitute power over life.

Following Foucault one can conclude that "power over life" involves more than one power apparatus. If so, biopolitics may be conceptualized as a strategic field consisting of multiple power apparatuses crosscut by contestations pertaining to vital phenomena. In the present work, biopolitics is understood to implicate many powers, not solely "biopower". Distinct from the multiple forms of power that traverse biopolitics is the specific power apparatus in dealing with the vital events of population. Again following Foucault's analysis, I make a sharp distinction between power over population and power over individual bodies, and for the sake of clarity will designate as "population power" the apparatus Foucault variously refers to as "biopolitics" and "biopower". The naming of these concepts is of course less significant than the reasoning that supports them, and I can see an argument for a concept unifying vital powers over the population and individual, as do Rabinow and Rose with the concept of "biopower", but that move would introduce differing power geneaology apparatuses into the vital events, a complexity that would need to be acknowledged and explored. At this point in the genealogy of the health function, the asymmetry between power over population health and power over individual health needs greater emphasis, particularly in the study of risk governance, which is all too often conflated with the self-governance of the neoliberal subject, a move that occludes risk at the level of population, epidemiology and public health.

My reasoning with respect to the analysis of population power as a single power apparatus and biopolitics as a strategic field of multiple power apparatuses flows in part from the empirical site investigated in the present book. The institution of a perinatal threshold for the living subject fell on the terrain of population power: interventions in vital events pertinent to population. But, since attempts to reduce perinatal mortality and morbidity occurred in clinical practice through risk assessment applied to individual bodies, the administration of life at the threshold of the living subject played out over the bodies of individuals. I have thus configured biopolitics to include population power, discipline, and other powers such as sovereignty.

Because the history of the present forms my main methodological point of departure, the study of biopolitics is pursued in relation to rare, truthful discourses. Historians of the present investigate a specific range of practices that are "more or less regulated, more or less deliberate, more or less finalized ways of doing things" (Foucault 1998: 463); its theoretical object is thus neither common knowledge nor experience (Foucault 1972: 220; Dreyfus and Rabinow 1983: 47–48). These discursive practices are normed by truth, that is, they interrogate their own internal statements according to a division between the true and the false, constantly expelling the false. The existence of "common knowledge" and experience is not denied, but simply remains outside the scope of inquiry to its gain and its cost. Rare discourses are studied with respect both to the procedures that form and circulate statements (archaeology) and to the powers that elicit, support and extend truthful statements (genealogy) (Foucault 1997e: 132). The daily life of peasants or workers thus does not constitute the subject of research, which instead consists of those knowledges typifying regions of truthful practice such as governance, disciplines and scientific knowledges, their conditions of existence, transformations and displacements.

Given that the history of the present constructs its theoretical object as the analysis of rare, truthful discourses imbricated in power relations, "governance" has a specialist usage differing from its everyday and social scientific senses as the party in political power or the state. Foucault's lecture, "Governmentality" (Foucault 1997a) traced the idea of the "art of government" to the sixteenth-century anti-Machiavellian commentaries on *The Prince*. The significance of these commentaries lay, he argued, in forming a style of reasoning for the state that separated the interests of the Prince from those of his subjects. In the anti-Machiavellian literature, the concept "reason of state" proposed a new kind of rule. Turning away from inquiries into the Prince and his territory, reason of state considered the subjects of rule as a *population* that had characteristics of significance to rule. Governance was to be based on the immanent characteristics of subjects viewed as a population. The invention of governance marked a form of political reasoning that aimed to govern "the complex of men and things" so as to maximize wealth, health and resources (Foucault 1997a: 208). In the art of governance, the *oikos* as kin-based farm was inserted on to the terrain of the state, constituting economy as a knowledge and goal of the state.

Among historians of the present, governance has come to be understood as a truthful form of reasoning associated with know-how for optimizing populations in terms of strategic goals such as wealth or health (Gordon 1991; Rose 1999). Studies of governance trace the replies given by expertise to questions they have posed regarding whom/what they govern, how to govern (including limits to governance), to what ends, and who they take themselves to be (Rose and Miller

1992: 178–179). This analytic of governance differs markedly from the theory of the state, a subject that has inspired commentary and debate (Curtis 1995; Rose and Miller 1992; Miller and Rose 1995). The focus of governmental studies falls on the technical reasoning of expertise inside and outside the state, extending across multiple sites from the governance of factories to self-governance. Unlike the theory of the state, investigations of governance assume no single, central locus of power. The analysis of governance opens to truthful discourses and their effects. It thus does not privilege the kind of institutional analysis associated with the theory of the state, which constitutes the state as a collective institutional actor formed from social relations and interests. Nor is the literature on governance concerned with the questions of legitimacy and authority that characterize normative political theory. Governance is a mobile political technology that passes below, through and across the institutional and territorial divisions of the state. While the theory of the state typically thematizes economies of control, domination and emancipation, analytics of governance treat cruelty and violence as an effect of relentlessly optimizing human populations.

Governance takes population as its main object, struggling perpetually to optimize it. As the sociologist Bruce Curtis (2001: 4) has emphasized, population is an abstraction rather than an empirically existing object. Population is the product of a governmental vision that treats political subjects as identical with respect to birth, death or other criteria. The statistics of population are gathered during an administratively defined time (such as the year) and for an administratively defined space (such as the territory of the nation) (Curtis 2001: 26). Curtis has argued that population as an abstraction involving the enumeration of all humans in a territory did not exist in France until the status differences of the Ancien Régime had been levelled at the close of the eighteenth century, making it possible to conceive of political subjects as sharing abstract equivalences (Curtis 2002: 529). In our present, perinatal mortality appears as a category in local, regional, national and global statistical systems, governed as part of the vital statistics of population. In counting perinatal mortality, stillbirths and neonatal deaths are rendered equivalent for the purposes of an administrative space – the hospital, region, nation or globe – and an administrative time that begins at twenty-eight weeks' gestation and ends seven days after birth.

The inclusion of stillbirths in perinatal mortality statistics constructs stillbirths as subjects in a population and as preventable deaths for the purposes of governing through health. Since the latter half of the nineteenth century the fetus/unborn has been regarded as "viable" during the last weeks of pregnancy, that is, as capable of living independently of the mother if a birth were to occur. Jeffrey Baker (1996: 22) reports that in the latter half of the nineteenth century no Western country placed the legal limit of viability under 180 days (just under twenty-six weeks). The concept of viability in turn depended on the configuration of the unborn as a biological fetus undergoing a sequence of developmental stages, a conception of the unborn that did not exist prior to the late eighteenth century (Duden 2002b: 37-39; Duden 1999: 18–24). The invention of the unborn as a fetus in Soemmerring's *Icones embryonum humanorum* (1799) (Enke 2002) coincides with the formation of modern biology (Foucault 1989: 286–304). By the first half of the nineteenth century in medical discourse, the unborn came to be configured as a fetus undergoing biological development during pregnancy, progressively coming closer to viability.

In the early modern period, "quickening" marked the moment when a woman sensed a stirring within her; to be "quick with child" was to sense the child as living/alive and to know herself as pregnant. Barbara Duden (2002a: 58–60; 2000) has termed this early modern perception of a hidden presence that had not yet manifested itself the "*non-dum*" (Latin meaning not yet). It was only at birth that the *non-dum* was revealed as a child or as "false fruit": flesh of differing kinds that was neither alive nor subjectified. Only after the birth threshold had been traversed did the uncertain knowledge of bodily interiors cohere into the certainty of a child, a living subject.

When from the 1950s the vital statistics of population came to include measures of perinatal mortality, the fetus after viability was implicitly postulated as a living subject whose death was to count and whose health was to be preserved. This subject can be called living because it was regarded as capable of dying, with its death multiply consequential for health governance: recorded in mortality statistics, subjected to perinatal death reviews, and mourned as perinatal loss. This living substance is a subject in the sense of being an effect of "forces, practices, and relations that strive or operate to render human being into diverse subject forms, capable of taking themselves as the subjects of their own and others' practices upon them" (Rose 1996: 171). The living subject prior to and during birth forms the lower limit of subjectification, being culturally understood as capable of being acted upon but not as capable of managing its own conduct.

With a number of exceptions (Hunt and Wickham 1994; Hunter 1994; Rose 1985) the history of the present has tended to examine single assemblages[6] – the school *or* the prison *or* the hospital – and one form of power – discipline *or* pastoral power *or* governance. This was the case with two previous book-length studies of pregnancy and childbirth affiliated with the history of the present: Jana Sawicki's *Disciplining Foucault* (1991) and William Arney's *Power and the Profession of Obstetrics* (1982), both of which centred on the concept of discipline. Sawicki argued that "new reproductive technologies" operate not by depriving women of reproductive power, but through a disciplinary incitement of desire, an argument that would appear to conflate the concept of discipline with that of positive power, that is power that operates through optimizing, inciting and making happen rather than prohibiting, oppressing and shutting down. Positive power characterizes the exercise of all forms of power other than sovereignty in foucauldian work, extending, for instance, to governance and security, and thus there is no *prima facie* reason to presume that recent developments in the governance of pregnancy and birth are confined to discipline. Arney's book also drew on the concept of discipline, interpreting the place of pregnancy care within obstetrics since the end of World War II as a disciplinary extension of surveillance outward from the hospital and the formation of new, continuous monitoring techniques (such as electronic monitoring of fetal heart rate during birth) for women in prenatal care and childbirth. The epidemiological forms of surveillance that stretched outward from the hospital are treated by Arney as a disciplinary panopticism, and monitoring as the application of disciplinary norms to the patient in care. As I will show below, the posthumous publication of Foucault's 1975–1976 lectures at the Collège de France (Foucault 2003 [1997]), first accessible long after the publication of Arney's 1982 book, suggests that a distinction be made between the epidemiological surveillance of population and the discipline of individual bodies as involving fundamentally different kinds of power. Yet *Power and the Profession of Obstetrics* remains the richest work produced by an historian of the present about the genealogy of pregnancy and childbirth, a fundamental historical and analytic contribution to the understanding of the post-1945 period which I draw on in Chapters 2 and 3 of this book.

The hold of discipline on social science has been tight, a generation transfixed by the eye of the panopticon, such was the power of that arresting image. The threshold of the living subject has been rendered knowable and manageable through diverse forms of power, only one of which is discipline. Disciplinary regimes act on the bodies of individuals, subjecting each to training in order to optimize their utility. In the present work, discipline has a restricted presence, appearing at two points: (1) when women are incited to be responsible managers of their own risk factors, and (2) when, refusing this form of self-governance, women are ordered into involuntary medical treatment. Risk is a technology of both security and discipline, and it was through the securitizing of population that the perinatal threshold was launched. Security acts on populations rather than individuals, anticipating the likelihood of future events which are evaluated in terms of comparative costs and profits. Techniques of security plan courses of action by calculating the mean and standard deviations rather than making a binary distinction between the permitted and the forbidden (Gordon 1991: 20; Dean 1999a: 116–118). When security methods are applied to vital phenomena, security takes the form of population power; both security and its subtype, population power, conceive their terrains of action as the random events that appear patterned only when analysed over large numbers. People die, but populations have mortality.

Prenatal risk assessment is used in clinical practice for the care of individuals. It thus has a double form belonging to both discipline and security, acting as a relay between the epidemiological norms established through the analysis of populations and the disciplinary norms of clinical practice. However, as I will show in the next section, the application of risk in clinical practice is not fully coherent, a point that has been repeatedly established and forgotten in the health sciences. Prenatal risk assessment continues to be used in health care, an effect of what might be called a will to health.

Legal regimes accompanied discipline and population power in the problematization of the threshold of the living subject. A long series of cases emerged in Canadian child welfare law and tort law that involved medical evidence of prenatal risk. These cases were typical of contemporary Euroamerican legal regimes in combining the normative power of prohibition with the positive power of governance. The initial critical reception of *Discipline and Punish* (1979 [1975]), Foucault's main writing on law, understood him to have argued that law was purely negative and juridical in contrast to the positive power of discipline. In that book, Foucault contrasted law, particularly juridical power as a right of seizure, with another form of power that acted to optimize the bodies of individuals through continuous observation, record-keeping, normalization and examination: disciplinary power. Foucault repeatedly used discipline as a contrast concept to sovereignty and law, which gave rise to the reading that discipline replaced law and sovereignty. More recent work among historians of the present has realized that this reading of *Discipline and Punish* was inaccurate; Foucault did not reduce law to "the juridical", instead arguing that since the late eighteenth century, law itself had combined juridical with other forms of expertise, notably discipline and security (Ewald 1990; Hunt and Wickham 1994; Rose and Valverde 1998). It is to this line of analysis that the present investigation of the perinatal threshold belongs.

Case law pertaining to the perinatal threshold was politically contested in relation to encroachments on women's liberty rights. Liberal governance has been key to international reproductive politics in the late twentieth century, and it too is not reducible to discipline. Liberalism[7] is a form of political reasoning that presupposes the existence of spheres – market and family being the classic examples – that possess immanent capacities for self-ordering. It configures these terrains as harmed if they are governed directly, although their functioning may be indirectly assisted through, for instance, education training subjects to govern themselves in accordance with their freedom. The direct governance of individuals and collectivities has been the subject of persistent contestation within liberalism, part of how it enables and directs its own self-critique.[8] When law aligns with liberal governance, it secures citizens as autonomous subjects with civil rights and freedoms guaranteed by the democratic sovereign.

Security/population power, discipline, law, liberal governance: problematizing the threshold of the living subject was accomplished through the concerting of multiple powers. It would as well seem that biopolitics contains non-veridical forms of contestation – claims-making that is not normed by the distinction between the true and the false – as sovereignty comprises a constituent of biopolitics.[9] Sovereign power enacts a symbolic register that does not conform to the logic of veridical discourses such as discipline (Singer and Weir forthcoming). Sovereign power includes the right to make and unmake law, to suspend law as such, to construct collective representations of the people/nation, to decide friend and enemy at the limits of the collective, to make a common, public world beyond the level of the purely local (Singer and Weir forthcoming).

Sovereignty is an achievement of symbolic language, constituting that which it represents in a manner that breaks with the realist constraints of the truthful discourses that comprise discipline and governance, which are based on the postulate of representing a plane of preexisting reality rather than a reality constituted through its representation.

When Canadian courts considered whether to dissolve the birth threshold at law, contemplating whether the "child in need of protection" (a category in statutory law) could be applied prior to birth and whether the centuries-old concept of the "child *en ventre sa mère*" referred to a fetus that could suffer injuries, the judiciary was arguably using its properly sovereign power to adjudicate who could be recognized as a legal person and what were the boundaries of the court's protection. The case law problematizing the threshold of the living subject in part involved law's symbolic, sovereign power to constitute that which it represents.

The history of the threshold of the living subject, and biopolitics more generally, may be analysed as occurring at the unstable conjunction of several interacting forms of power: what I call the *power field* of biopolitics. This concept is in dialogue with Foucault's famous remarks (1997a: 219) about the coexistence of differing powers. He cautioned that powers do not simply displace each other in historical sequence; it is a misconception to frame the historical analysis of power in terms of a "replacement of a society of sovereignty by a disciplinary society and the subsequent replacement of a disciplinary society by a society of government; in reality one has a triangle: sovereignty–discipline–government" (Foucault 1997a: 219). *Power field* names that triangle: an array of powers bound together in strategic relations to each other. But there is no theoretical requirement that powers be limited to groups of three.

An historical epoch has no single "diagram" in the sense suggested by Deleuze (1988), who mistakenly took Foucault to argue that epochs are usually characterized by the dominance of a single form of power such as sovereignty or discipline.[10] Instead, an historical epoch is characterized by multiple coexisting powers. Strategic fields of contest such as biopolitics may also be traversed by multiple powers. The power field of biopolitics implicated in the problematization of the threshold of the living subject in the twentieth century was characterized by an agonistic conjunction of four powers: security of population (population power), discipline of individual bodies, the sovereign power of law, and the liberal/authoritarian stylization of governance. These four powers might be grouped as two alternative strategies: security–law–liberal governance and discipline–law–authoritarian governance. Disciplinary confinement and correction would be activated when the subject failed to comply with the incitements of security. For the disciplinary triad to be invoked, security must be invested with life-and-death significance such that the function of modern power to implant life is perceived to be compromised. Under these conditions, which include infectious disease outbreak and epidemic, liberal freedoms may be suspended.

The power field of biopolitics gives rise to impermanent stabilizations among its component powers. Security absolutized through the suspension of law, without regard to liberal governance, may take the form of genocide. But security may fall within the warrant of law, though outside the constraints of liberal governance. The attempts by child welfare agencies to gain jurisdiction over the unborn/fetus constituted an instance of this second possibility: the alignment of law with discipline whereby subjects are stripped of freedoms due to their not conducting themselves in accordance with security. Child welfare authorities asked the courts to order pregnant women into detention and mandatory treatment for failure to act in conformity with perinatal risk aversion. Given that modern politics is accountable for the health of populations, the persistent question arises as to whether those who will not govern themselves in accordance with health might be compelled to do so through law. The case and statutory law about involuntary medical treatment – e.g. coerced blood transfusions or the forced removal of organs from one family member for the sake of another family member – occurs at the unstable conjunction of liberal governance with the security of populations. Law, security and liberal governance are constantly thrown into and out of alignment in biopolitics, with legally mandated discipline used as a secondary strategy in liberal regimes.

The history of the present has for the most part dealt with discursive processes that act to incite health rather than to compel it, with the case of genocide offered as the antithesis of incitement. But, as Mitchell Dean (1999b: 131–148) has demonstrated, authoritarianism in biopolitics comprises a variety of practices less spectacular and bloody than genocide. Court-ordered detention and mandatory treatment of pregnant women for the sake of fetal health illustrate biopolitics gone authoritarian. Medically based prenatal risk techniques that pushed the threshold of the living subject prior to birth were the condition for this form of intervention. As physicians, nurses and social workers came to give expert evidence that fetal health was at risk from the conduct of pregnant women, the question arose as to how the unborn were to be conceptualized legally and whether the courts could compel women to conduct themselves so as to preserve fetal health. The end of the problematization of the threshold of the living subject occurred with decisions by the Supreme Court of Canada that ultimately confirmed the birth threshold and the liberal governance of women. There are many other jurisdictions, notably the United States of America, in which this kind of determinative judgement at the highest judicial level has not occurred.

Risk at the threshold of the living subject

The techniques of the threshold secure its dependability. The threshold of the living subject has come to be thickly governed through risk techniques. In the second half of the twentieth century and continuing into the twenty-first, risk techniques attached pregnancy and birth to the perinatal threshold. Physicians chart "prenatal" health during pregnancy through a grid of risk factors. Courts award damages for the care of injured children whose health was placed at risk *in utero* from motor vehicle accidents. Warnings abound on cigarette packages, posters in bars, and television commercials about risk to the unborn/fetus. Amid this proliferation of practices and representations one might wonder how it was that pregnancy came to be so densely connected to risk.

Historians of the present analyse risk as a "technology of governance". Governance forms at once a way of reasoning (a governmental rationality) and a way of making reasoning operable (a governmental technology). It is eminently practical and experimental (Miller and Rose 1990: 8). Technologies of government are about the "how" side of "know-how", an instrumental rationality presupposing that obstacles can be fashioned into solvable problems.

Risk orients governmental practice to the provision of security: a felicific calculus of utilitarian ends. Enabling prevention, risk governance attempts to arrest future harm before it exists by selecting options for action based on the calculation of probabilities. Taken in this sense, risk is not equivalent to harm or danger: it is a way of reasoning and acting that anticipates future harm or danger. This way of conceptualizing risk is quite distinct from other analyses of risk in the social sciences, some making risk into an objective category of the real, others understanding risk as a category of common culture. Risk governance consists of rationalized schemas that estimate the likelihood of unwanted future harms. Governing pregnancy through risk at the threshold of the living subject mobilizes heterogeneous risk techniques – indeed, displays the heterogeneity of risk governance – that share the general form of risk as calculative, preventive and oriented to future security.

Risk techniques and risk effects are diverse and socially pervasive in our present. The heterogeneity of risk techniques includes the social insurance technologies of the welfare state (Ewald 1986), harm reduction in the management of addiction (O'Malley 1999), the organization of prison populations and parole (Feeley and Simon 1994; Kemshall 1998; Simon 1988 and 1993), and health promotion campaigns (Bunton, Nettleton and Burrows 1995; Gabe 1995; Petersen 1996). Yet emphasis on the pervasiveness of risk should not be mistaken for the triumphant march of risk through modernity. The "risk society" does not exist because the present is not ordered according to any single principle; biopolitics is not reducible to risk.

I distinguish four distinct risk techniques active in the formation of the perinatal threshold: epidemiological, clinical, actuarial, and legal risk. My interest here lies not in producing an exhaustive typology of risk, but rather in foregrounding the analytic specificity of risk techniques and their varying effects in health, insurance and law.

Within medicine, the concept of risk is associated with epidemiology, the quantitative study of the occurrence and distribution of disease in human populations. In epidemiology, risk is part of a class known as "incidence measures": the study of new cases of a disease/condition appearing within a specific time interval. Risk has a precise methodological sense in cohort studies that investigate the relation of what are called "exposures" to "outcomes" in a study population. Epidemiological risk is calculated over populations, with the study population conventionally formatted in a 2 x 2 table (see Figure 1.1).

The 2 x 2 table provides a grid of intelligibility for risk in an epidemiological sense: a combination of four cells (a, b, c, d) containing the numbers of the study population with and without the exposure (e.g. smoking) at the beginning of the observation period, crosscut by those who did and did not have the outcome in question (e.g. lung cancer). One can see that there are a number of risk measures based on differing combinations of the cells in Figure 1.1, some of the most commonly used being relative risk, attributable risk, and population attributable risk (for formulae see Figure 1.1).

Where epidemiology studies populations, clinical medicine involves the diagnosis and treatment of patients. Disease for clinical purposes is conceived as immanent in the body of the patient, taking variable courses through the individual body, courses that the clinical gaze must discern. *Clinical risk*, a concept first introduced by Sandra Gifford (1986) and used by her contrastive to

	Exposure		
Outcome	*Yes*	*No*	*Total*
Yes	a	b	a + b
No	c	d	c + d
Total	a + c	b + d	a + b + c + d

Risk measures

Relative risk

- ratio of risk in exposed to non-exposed: a/a + c ÷ b/b + d
- indicates how much more likely it is that the exposed develop a disease/condition than the non-exposed

Attributable risk

- risk in exposed *minus* risk in non-exposed: a/a + c – b/b + d
- additional risk in the exposed; a measure of a causal relationship between exposure and outcome

Population attributable risk

- risk in study population *minus* risk in non-exposed: (a + b)/(a + b + c + d) – (b/b + d)
- additional risk in the total study population associated with the exposure; reflects a causal relationship between exposure and outcome

Figure 1.1 Calculating epidemiological risk

epidemiological risk, indicates the application of an epidemiological knowledge of risk in population to patients in care. When epidemiological risk factors are inserted into clinical practice a transformation occurs. In the clinical setting risk factors are read as equivalent to "objective clinical signs of disease" (Gifford 1986: 222) under conditions of uncertainty.[11] Where in epidemiology risk from an exposure is a quantitative concept, calculable over populations, risk from an exposure to an individual in clinical care is not calculable. In clinical practice, it is uncertain whether a patient with an exposure should be assigned to cell a or c in Figure 1.1. Since the outcome is uncertain in the sense of being incalculable, clinical risk is not strictly speaking risk, but, as I will show, a composite form of incommensurable judgements.

Within the history of the present, risk judgement and the diagnostic judgement of the normal and the pathological have been treated as analytically distinct and historically sequential. David Armstrong (1995: 395–397) has argued that risk constituted a new kind of clinical judgement when it was introduced in the 1950s, diverging from earlier forms of medical reasoning that had been based on distinguishing the normal from the pathological. Bodies "at risk" are neither sick nor healthy. Risk and diagnostic judgements are also dissimilar in their constructions of time and space. Risk judgements orient to the future, diagnosis to the present. Diagnosis refers to processes occurring in the depths of the body, whereas risk exposures may be conceptualized as located inside or outside the body, the latter in "environment" and "lifestyle" (Armstrong 1995: 400–401).

Although judgements of risk and judgements of the normal and the pathological are analytically distinct, I am concerned with their co-occurrence and interaction. In Chapter 2, I show that risk in the care of individual patients is characterized by an interesting and messy conjunction of the two judgements. Indeed, all clinical interventions from test results to diagnoses are sometimes termed risk, which thus comes to be coextensive with any and all clinical interventions. Clinical risk comprises an unstable amalgam of incompatible forms of reasoning.

The third technique considered here, *actuarial risk*, is part of insurance. Actuarial risk insures a capital against the probability of its loss calculated over an interval of time (Ewald 1991 and 2002). It operates by spreading the cost of anticipated losses over a list of subscribers, the members of a risk pool.

And finally, I use the phrase *legal risk* to indicate the imbrication of risk reasoning in legal discourse. In the case law analysed in Chapters 4 and 5, risk enters legal discourse through medical evidence and explicit mention in statute law. Legal risk indicates as well the entry of risk into legal reasoning at the level of summaries of fact in judgements and legal concepts such as the standard of conduct. Legal risk may be compatible or incompatible with liberal governance, but, because it seeks to prevent that which has not occurred, it has presented recurring dilemmas for governing through freedom.

Although epidemiological, clinical, actuarial and legal risk techniques differ in their analytic forms, their joint use has introduced new intensities of care for the living subject at the threshold of its arrival. Taken together, the four risk techniques (epidemiological, clinical, actuarial and legal) have acted to conserve the health of the unborn for the sake of child health, thus instituting a perinatal threshold for the living subject.

During the 1950s, medical researchers applied epidemiological risk to the study of perinatal mortality, identifying long lists of risk factors. This research constituted perinatal mortality as preventable through the management of risk factors in clinical care. I will show how the lists of perinatal risk factors were folded into a preexisting system of prenatal care and a grid of risk factors applied to the bodies of pregnant women – clinical risk.

Actuarial risk at the threshold of the living subject has a genealogy separate from epidemiology and health until the mid-twentieth century, its focus to that point having been insurance claims after fatal accidents or personal injuries. From the mid-nineteenth century, actuarial risk facilitated a new application of the child *en ventre sa mère* in fatal occupational accident statutes, giving a child born alive after a parent (typically a father) had died in a workplace accident the entitlement to receive survivor benefits. Roughly a century later children born alive with injuries from accidents that had occurred during pregnancy were first permitted to sue for damages in the tort of negligence. These legal actions had actuarial risk as a condition of existence; without compulsory accident insurance, there would rarely have been any point to launching a legal action.

Legal risk comprises the last of the risk forms through which the threshold of the living subject was problematized. Medical evidence pertaining to fetal risk entered tort law and child welfare law during the mid- to late twentieth century. In Canada during the last two decades of the twentieth century child welfare authorities attempted to enlarge their jurisdiction to the unborn/fetus, using medical evidence of epidemiological and clinical risk in arguing their cases, asking for judicial orders protecting fetal health from maternal risk factors such as alcohol use or glue sniffing. When legal risk was used to problematize the threshold of the living subject, it activated the sovereign power of law, symbolically to constitute the legal person, specifically the time at which the legal person might be said to first exist.

The antithesis of calculable risk is incalculable uncertainty (Knight 1921). Recent work among historians of the present has investigated complementarities between governing through risk and governing through uncertainty (Ewald 2002; Haggerty 2003; O'Malley 2003 and 2004). As Janice Richardson (2004: 104) has noted, Pat O'Malley's writings have been distinctive in examining the operation of risk and uncertainty within legal reasoning rather than externally at the level of insurance. O'Malley (2004; 2003; 2000) has examined the subjectification of the entrepreneur expected to pragmatically launch courses of action into incalculable futures, alert to possibilities for profit in the smallest and largest of processes. In the law of contract, O'Malley (2003: 236) argues, the legal subject is normatively anticipated to have a good knowledge of market conditions, but, as the future of the market cannot be predicted, the legal subject is expected to govern him/herself through practical foresight, that is, uncertainty. Other historians of the present (Ewald 2002: Ericson and Doyle 2004: 212–283) have examined the governance of "terrorism" and major disasters such as nuclear power plant explosions and the unintended release of genetically recombinant infectious pathogens. Through the precautionary principle, the governance of incalculable uncertainty attempts to pre-empt risk.

The governance of risk and uncertainty interact at the perinatal threshold, with risk continually and unstably reconfigured as uncertainty. Epidemiological risk enters uncertainty when applied to the bodies of patients in clinical risk. In child welfare litigation, epidemiological and clinical estimates of risk to fetal and child health from maternal conduct during pregnancy were used to justify extending the child in need of protection, a concept in statute law, to the unborn. But the application of risk concepts to the bodies of individuals is always technically uncertain; this is widely known in health expertise. In tort law the unborn child, conceptualized as a fetus in the evidence of expert witnesses, became a foreseeable class of persons that the responsible legal subject should anticipate as affected by his/her actions. Risk enters litigation in evidence, and becomes judicially reframed as a standard of conduct for the legal subject governing her/himself in uncertainty.

A form of technical reasoning, risk has been joined to differing discourses, institutions and forms of power. At the threshold of the living subject, risk techniques have been exercised in and across health and law as governmental "complexes": assemblings of discourses, practices, agents, authorities, textual forms, functions and forms of judgement.[12] "Complex" thus marks a rough boundary between health and law. Risk is a discursive practice which, when attached to the health complex or the legal complex, is configured according to the practices of that complex. The activities of risk within a particular complex thus cannot be explained by invariant attributes of risk. Clinical risk, for instance, was invented in the health complex as the application of risk factors to the bodies of individual patients. When clinical risk entered the legal complex through medical evidence about risks to fetal health, specific legal questions arose in differing branches of law. In child welfare law, it was asked if the statutory meaning of "child in need of protection" should encompass the fetus/unborn, while in the tort of negligence, the very admissibility of medical evidence was debated. Attaching perinatal risk information to these branches of law problematized the legal address of pregnancy and the demarcation of the legal person. Risk serves security/population power in ways specific to the complex in which it is enacted.

The emphasis placed here on the heterogeneity of risk techniques runs against the tendency to regard risk as a unity, a common feature of existing literatures on risk in the social sciences. Mary Douglas (1992), alive to the history of risk concepts and to comparative social analysis, traces the history of risk from its uses in mathematical calculations of probability in securing against losses to its contemporary sense of "danger" in political argument, a mechanism of blame and accountability. In Ulrich Beck's work, the "risk society" indicates a second phase within modernity where "the institutions of industrial society produce and legitimate hazards which they cannot control" (Beck 1995: 27). Beck treats risk as "hazards" that threaten humanity in our present, conceived as a particular historical stage: the "risk society". As Mitchell Dean (1999a: 181) has noted, Beck assumes the "uniformity" of risk, that it has "fundamentally the same characteristics in all spheres". Douglas and Beck each render risk as singular in meaning – either danger or hazard.

Until recently, historians of the present understood risk as solely "actuarial" in the sense described above: a means of spreading the costs of accidents across the members of a risk pool through insurance. The effects of actuarial risk were pervasive in the twentieth century, found in private and public insurance, including the social insurance schemes of the welfare state. In previous work (Weir 1996 and 1997), I was concerned to argue that actuarial risk constitutes only one type of risk technique. Another and quite distinct form of risk, what I variously called "clinical risk" or "epidemiological risk", has been present in health care (Weir 1996: 381–383). Clinical and epidemiological risk do not function as mechanisms of loss redistribution. They act on individuals and populations to maximize health.[13] While Dean (1999b: 131) may have overstated the point about the diversity of risk in arguing that "the significance of risk does not lie with risk *itself* but with what risk gets attached to", since risk is uniformly a technology of security, his comment draws attention to the differing attachments of risk in health, law, insurance and other complexes.

The much-read collection, *Embracing Risk* (Baker and Simon 2002), notes that, at the turn of the twenty-first century, individuals are incited to "embrace risk", with collective responsibility for risk discouraged. The Baker and Simon collection associates the trend to encourage individual risk-taking with the rise of "risk beyond insurance" after the demise of the welfare state. However, in the health complex, epidemiological and clinical risk were invented decades prior to and independent of neoliberal attempts to encourage people to be responsible risk managers of their own lives. Risk beyond insurance has a genealogy separate from the neoliberal encouragement of individual risk-taking and risk aversion.

Because risk techniques have been analysed with particular reference to liberalism and neoliberalism, risk itself has come to be associated with programmes of liberal governance. The presumption that all risk is actuarial and liberal/neoliberal is in the process of being rethought from a number of different directions. Pat O'Malley (1992, 1996, 1997, 1999, 2000, 2003, 2004), in an overlapping series of papers, has examined the history of risk and welfare together with current risk techniques in crime prevention. The combined outcome of these papers has been to show that risk techniques are "politically polyvalent" (O'Malley 1996): risk has no intrinsic political belonging. He has argued that risk has been remarkably adaptable, "incorporating diverse technical and moral configurations" (O'Malley 2000: 458).

Risk in health governance has been characterized by heterogeneous political belongings, operating in alliance with or rejection of liberal governance. At the perinatal threshold, risk acted to problematize the liberal governance of women, proposing fetal health risks be minimized by compelling women to comply with risk aversive practices through confinement and involuntary health treatments. As I will show, many of the key case law precedents involved Aboriginal women, who in the long colonial history of the Americas have been repeatedly positioned as subjects lacking the rational capacity for autonomy and freedom. The proposal to legally compel women to engage in conduct avoiding fetal risk was, however, rejected by the courts at the appropriate level.

The literature on risk and health has engaged solely with risk in health promotion. Risk governance in health is reduced to a logic that would enrol patients as active citizens in the management of their own health, making them responsible for their own risk reduction (Bunton, Nettleton and Burrows 1995; Gabe 1995; Polzer 2005; Rose and Novas 2004; Petersen 1996). Advanced liberalism, it is said, operates by "empowering" "communities" and persons. Thus risk governance of health is equated with risk governance of health promotion, and health promotion with neoliberalism.

I nowhere examine risk in health promotion. This is in part due to the fact that studies of health promotion and pregnancy have already appeared (Lupton 1999; Oaks 2000). It is also the case that locating prenatal risk governance in the practice of health promotion errs on several grounds. First, prenatal risk assessment was invented during the 1950s, entering widespread clinical practice by the late 1960s. Risk governance in health thus preceded advanced liberalism and occurred independently of it. Second, since the 1960s prenatal risk assessment in clinical care has operated separately from health promotion located in the public health sector. Clinical risk is sited within curative medicine, not public health. Third, risk factors found in the clinical care of pregnancy are far more extensive than the prenatal risk factors relevant to health promotion. Health promotion programmes have encouraged women to alter their "modifiable risk factors" through public health campaigns: to stop drinking, smoking, and using recreational drugs, to eat sincerely nutritious meals, and take folic acid. However, the risks present in prenatal risk assessment are diverse, and include those risk determinants such as diabetes and parity (number of previous births) that are not subject to modification through intervention in maternal conduct. Fourth, health promotion represents a model of public health that in Canada was displaced during 1994 by population health (Evans, Barer and Marmor 1994; Hayes and Dunn 1998), but risk-based prenatal care has continued unabated. Population health is concerned with the determinants of health in populations rather than in individuals or risk groups; it has been critical of the limitations of "lifestylism" found in health promotion. Conflating prenatal risk with health promotion fails on many grounds.

Current social science literature has been fascinated by the growth and massive presence of risk in contemporary governance. Expertise has been understood as proliferating risk, with debates internal to expertise assumed to take the form of how to make risk governance run better. Where the flow of risk has met with barriers, studies document these as the work of forces external to expertise. Rayna Rapp's (1999: 53–72) sensitive ethnographic work has shown the complexity of communicating and interpreting risk in clinical situations. Social science literatures have reported on popular struggles and perceptions of risk that act to discredit the purported neutrality of risk administrators (Michael 1992; Slovic 1987; Wynne 1995 and 1996). No doubt risk governance has expanded under neoliberalism, but this point has been argued in a one-sided manner without reference to the restrictions placed on risk by expertise itself. When risk attaches to the health or legal complexes, it interacts with prior forms of reasoning and techniques found there. These preexisting forms of reasoning bind with risk in complex ways; they facilitate risk governance, specify its attachment to a complex, or act to limit and reject it.

The study of risk governance at the threshold of the living subject shows the existence of divisions within expertise acting to both assist and restrict the flow of risk in the health and legal complexes. Standardized prenatal risk assessment has been criticized on many grounds by evidence-based medicine, among them the conflation of risk with diagnostic test results (a practice that inflates the reliability of risk assessment) as well as the iatrogenic effects of risk assessment. These critiques have been taken up within midwifery, which has allied itself with research-based critiques of obstetrics in order to justify its own alternative practices. Chapter 3 asks how midwives position themselves in relation to the risk techniques through which the perinatal threshold of the living subject was implemented. Using interview and archival sources, I investigate the reasoning midwives have employed to limit the use of risk judgements in prenatal care, showing that the goals of pregnancy care for midwifery are broader than lowering perinatal mortality and morbidity, which is the aim of risk-based care. Prenatal risk assessment is filled with medical and social relevances that concern midwives, a finding that resonates with a small research literature on midwifery and risk perception (Lane 1995; Murphy 1994; Oakley and Houd 1990; Saxell 1994). Midwives bring relevances to clinical practice that are absent from risk-based prenatal care: advocacy for pregnancy as a state of health and the stylization of the expertise–subject relations as collaborative. Seeking to create a new culture of birth, midwives defend the primacy of clinical reasoning over standardized prenatal risk assessment. Contemporary health governance thus contains styles of reasoning that exceed risk and that act to limit and contain risk reasoning.

Nor has law operated as a machine proliferating risk at the perinatal threshold. Expert medical evidence about prenatal risk moved the fetus/unborn from the health to the complex to law where it proceeded to unfix the birth threshold in the law of tort and child welfare law. However, the history of risk governance in these branches of law has been one of fitful expansion and contraction.

The ancient legal fiction (a principle of legal reasoning that marks a concept as a nonliteral, figurative departure from ordinary legal usage) of estate law, *qui in utero est*/the child *en ventre sa mère*, found a new application in twentieth-century tort law when the right of action to recover damages for prenatal injuries was first recognized in an international precedent set by the Supreme Court of Canada in *Montreal Tramways v. Léveillé* (1933).[14] The Court also created a precedent in allowing medical evidence to be admitted in claims for prenatal injuries. The admission of medical evidence permitted the legal fiction of the child *en ventre sa mère* to link with the biomedical concept of the fetus in tort law for the first time. Actions in negligence torts remained subject to the "live-birth" rule, the legal requirement that claims could only be brought on behalf of children born alive. But during the 1990s Canadian and United States courts heard arguments proposing that pregnant women as a class owed a duty of care to the unborn, an argument that would have overturned the live-birth rule and constructed a perinatal threshold for the living subject in the tort of negligence. This amounted to a demand for a fundamental revision of tort law, one that challenged the boundary between legal and medical discourse. It was a proposal of such magnitude that it was rejected to Canadian courts on appeal, an affirmation of the birth threshold.

Chapter 5 of this book examines a second line of legal cases that sought to displace the birth threshold, cases brought by Canadian child welfare authorities. Child welfare agencies launched these actions with the aim of securing the future health of children in their custody by preventing risks to fetal health during pregnancy. Research showed that children who were ill and disabled were more likely to remain permanent wards of the state than those in good health. Child protection services attempted to find some and any legal means of securing children's health after birth by acting on women during pregnancy and labour. But their mandate in statutory law was defined as "the child in need of protection", and that legal status had in practice been recognized to occur only at birth. In order to accomplish an extension of their jurisdiction to the unborn/fetus, child welfare authorities sought judicial authorization to detain pregnant women for involuntary medical treatment. Expert evidence by health care professionals was given as proof of fact to show that the health of children was jeopardized by risks to fetal health as the result of maternal conduct. Much, though not all, expert evidence took the form of risk estimates since interventions in fetal health have been pervasively organized through a risk-based regime of perinatal care. The end of the problematization of the threshold of the living subject in the child welfare litigation resulted in an affirmation of the birth threshold and the liberal governance of women, a defeat for risk-based arguments pertaining to fetal health.

When scientific expertise is used in litigation to establish fact and proof, scientific evidence is heard selectively in the courtroom. The process of establishing proof in litigation differs markedly from proof within science; law, as Sheila Jasanoff had argued, "constructs expertise" (Jasanoff 1995: 42).

Jasanoff (1995: 128) has shown how, in United States tort law dealing with ecological disasters – "toxic torts" – the judicial system showed a "preference for observable, empirical proofs" over evidence from population-based studies, whether animal studies or epidemiological ones. Judicial fact-finding preferred the accounts of physicians who had examined patients clinically to epidemiological studies (Jasanoff 1995: 123). In the tort and child welfare case law about the unborn that I examine here, the judicial system repeatedly dealt with the question about the very admissibility of medical evidence, both clinical and epidemiological; the boundary between epidemiological and clinical evidence was not sharply drawn as the unborn were made knowable in clinical practice from the mid-twentieth century through risk judgements. Medical risk evidence, clinical and epidemiological, was used to link the born with the unborn across birth, but the Canadian judicial system did not proceed unproblematically to authorize clinical knowledge and its stories of individual suffering over more abstract epidemiological accounts of outcomes and exposures. These cases involved legal precedents in tort and child welfare law such as whether to permit a right of recovery for prenatal injuries and whether to encompass the unborn within the meaning of a child in need of protection under provincial statutory law. At stake in many of these cases was whether evidence was legally admissible or relevant.

Much scholarship exists regarding maternal–fetal conflict at law in many jurisdictions (Daniels 1993; Dawson 1990; Morgan 1989; Petchesky 1987). Legal regimes have been uncertain about its relation to the maternal–fetal distinction; a long problematization of the legal address of pregnancy has occurred. I call attention to the appearance and significance of medical evidence about fetal health risks in child welfare and tort litigation. The concept of prenatal and perinatal health risks acted to form a tie between the born and the unborn, establishing continuities of health across birth, rather than constituting the fetus as a substantive being distinct from the child after birth. Risk evidence enabled proposals for a perinatal threshold to be made in case law, potentially extending into law a logic of conserving population across birth. The case law examined here shows a logic of debating the consequences of encompassing the unborn in the legal status of the child after birth in order to preserve the health of the child; risk reasoning enables the conceptualization of health and temporal continuities across the birth threshold.

Prenatal risk assessment was invented within the health complex as a technique to reduce the mortality and morbidity of the fetus and the neonate during the perinatal period, an interval of time extending from the last weeks of pregnancy through the first week of birth (twenty-eight days in some countries). There have been numerous social scientific studies of the neonatal period, particularly neonatal intensive care (Anspach 1993; Frohock 1986; Heimer and Staffen 1998). The prenatal aspect of the perinatal (formally the twenty-eighth to the fortieth week of pregnancy) is of particular interest here because it created a new division in clinical care, dividing the time of the fetus/unborn from the time

earlier forms of prenatal care had been about maternal health, and prenatal education about the health of infants, risk assessment was novel in its attention to the health of the fetus. With the concept of the perinatal interval and the associated technique of prenatal risk assessment, the biopolitics of population transcended birth and came to include the unborn conceptualized as fetus. Pregnancy itself became a time for routinely conserving and optimizing population prior to birth, subject to new intensities of care.

Minimizing perinatal mortality rates constituted something of a paradox: the prevention of death prior to birth. The paradox existed in cultures where birth had marked the simultaneous conferring of personhood and individuality on the newborn (Conklin and Morgan 1996); prior to birth neither persons nor individuals existed. How then could the death of someone not born be possible? Thus, while the aim of governing perinatal mortality was clear (reducing the perinatal mortality rate), what/who was being conserved was obscure, particularly in legal regimes, where the new subject of mortality set in motion a series of quandaries.

When the birth threshold of the living subject came unfixed, so too did the relations of before and after, inside and outside, that had been held apart by birth. The birth threshold was no longer a dependable between. Moving the threshold before birth attached the arrival of the living subject to the inside of the maternal body during pregnancy. How was this strange threshold, the perinatal one, to be governed? Lived?[17] How was risk to fetal health to be authorized at law? Should pregnant women be compelled to be healthy for the security of future populations? Was someone whose behaviour had placed a fetus at health risk to be held liable for damages should the child be born alive with injuries? Should insurance coverage be construed as extending prior to birth? Did personhood at law exist prior to birth? Court orders for women to stop activities that increase risk to the unborn/fetus, on pain of incarceration and coerced medical treatment, offered one solution to this problematization.

The introduction of the perinatal threshold into law created a crisis of intelligibility around the legal address of pregnancy and the beginnings of the legal person, catching the biopolitics of population in a recurring dilemma of liberal governance. Pregnant women could be positioned as citizens with legal and political rights as members of the polity (what was called the "city–citizen" game by Foucault [1997c: 311], who saw it as dating from the polis) or they could be positioned as the subjects of pastoral care: living beings owing obedience to the pastor and in turn receiving care (the "shepherd–flock" game dating to early Christianity [Foucault 1997c: 300–311]).[18] This conflict is hardly unique to pregnancy; it ran through the welfare state, where citizens were positioned unstably as citizens with rights and freedoms and as subjects of state solicitude in security measures (Dean 1999a: 96). The two positions have run into repeated conflict in the history of biopolitics, where the subjects of governance individually and collectively may be recognized as living beings/units of population whose suffering in health is to be alleviated, and as citizens with rights and freedoms. Those who do not comply with pastoral security may be consigned to disciplinary correction.

The pastoral positioning of the subject of health care – the subject as the recipient of health care for his/her own good or the good of a collectivity – represents a form of authoritarianism intrinsic to modern biopolitics. This vital authoritarianism has been questioned and limited by a growing hedge of rights and freedoms around the subject of health. Legal risk in pregnancy forms one iteration of the pastoral positioning of the subject in health care, a vital authoritarianism locked in conflict during the late twentieth century with a project to govern women as liberal citizens according to their freedom. By the end of the twentieth century the birth threshold had been reaffirmed at law in Canada and the United Kingdom, but the problematizing of the threshold of the living subject has continued unresolved in other legal jurisdictions, notably in the United States of America. The threshold of the living subject had again become dependable at law in a small number of nation-states, but it remained unsettled internationally in the health complex, where the perinatal threshold has continued to flourish, and from the health complex into common culture.

2 A genealogy of perinatal mortality

Perinatal [Fr. peri, around, + L. natalis, birth] concerning the period beginning after the 28th week of pregnancy and ending 28 days after birth.

(*Taber's Cyclopedic Medical Dictionary* 1997, Philadelphia: F.A. Davis, 1997, 18th edn)

Perinatal Care: from the 28th week of pregnancy through 7 days after birth; care of fetus and neonate: differentiate from prenatal care and postnatal care which are of the mother.

(Annotation, MESH (Medical Subject Headings), National Library of Medicine, National Institutes of Health, Bethesda, Maryland, http://igm.nlm.nih.gov, last accessed October 2005)

Now that maternal mortality is so low, the death rate amongst babies is regarded as a more satisfactory and more sensitive index of the efficiency of maternity services. The term perinatal mortality, defined as the number of stillbirths and deaths in the first week after birth per 1000 total births, is being used more in this connection.

(Sir Dugald Baird 1960, "The Evolution of Modern Obstetrics", *Lancet*, 2 (Sept. 10): 559)

Perinatal. National vital statistics report on perinatal mortality annually. Routine risk assessment of pregnant women is accomplished in the name of lowering perinatal mortality and morbidity. Chart reviews of perinatal deaths are performed in hospitals as a requirement for certification of obstetrical departments. Perinatology is a medical subspecialty in pediatrics. Pregnancy and childbirth in our present have come to be thickly and meticulously governed in the curious name of the perinatal.

Medical textbooks conventionally frame the history of pregnancy in the twentieth century as the progressive conquest of death measured in a historical sequence leading from infant to maternal to neonatal mortality, ending in our present: an age of perinatal mortality (see Dugald Baird's comment above). The reduction of the perinatal mortality rate is the limit against which contemporary health care has chosen to strive in maternal and child health, an antagonist appropriate to an age when infant and maternal mortality in advanced capitalist nations has fallen to what is called an "irreducible minimum". The medico-administrative categories of maternal, infant and neonatal mortality repeat social statuses found in common culture: mothers, infants and newborns. However, "perinates" do not exist as a subject in either health or common culture. "Perinatal" constitutes a medical neologism operating at the threshold of the living subject, naming a continuous temporal interval before, during, and after birth. The time of the perinatal constructs a particular subject/object for health intervention; the perinatal pertains to the fetus (after viability – typically judged as the twenty-eighth week of pregnancy in the early to mid-twentieth century) and the neonate (the duration of which is still debated internationally, taken as either seven or twenty-eight days after birth). The continuity of the perinatal interval depends on a conceptualization of commonalities in the bodily substance of the unborn and newborns; the unborn late in pregnancy and newborns are thought to die of the same causes. "Perinatal" fashions temporal and bodily continuities across the divide of birth.

I will trace a short genealogy of how the "perinatal" was invented and its subsequent attachment to risk techniques intended to reduce the perinatal mortality rate. The genealogy of the perinatal has three levels of analysis: the formation of "perinatal" as a concept (epistemological), the invention of risk-based prenatal care (technical), and the formation of the perinatal mortality rate as a target for the health complex (governmental). The linkages across concept–technique–governance co-ordinated the objectives of the health complex with the clinical care of individuals: the security of each and all.

The genealogy of the perinatal threshold offered here differs from medical histories of perinatology that claim an unbroken lineage from Hippocrates (Dunn 1993), vacuuming up the concepts, diagnostic techniques and therapies devised from the 1950s to our present in a happy retrospective synthesis of breakthroughs in perinatology. Such a developmentalist perspective overrides the difficult question of how perinatology itself came into existence. The heterogeneous activities made possible by the perinatal as a temporal concept were not driven by perinatology as a medical specialty. Rather, perinatology constituted itself as a subspecialty within pediatrics decades after the concept of the perinatal had been formulated and techniques for the reduction of perinatal mortality consolidated.[1] Perinatology arrives on the scene of the perinatal *post festum*.

The perinatal interval was first conceptualized within social medicine and pathological anatomy during the period 1920–1950 after analyses of vital statistics had shown that infant mortality rates in Western Europe and North America had fallen, but mortality among newborns had not. In the time between World Wars I and II, deaths among newborns came to be attributed to the same causes as stillbirths. Thus "fetal and neonatal" were joined together in a concept of "fetal and neonatal mortality" during the 1930s and 1940s, but "perinatal" did not become standard terminology until the mid-1950s, drawing on an earlier neologism formulated in mid-1930s Austria. Once the unborn were considered very like newborns in national regimes of vital statistics, their mortality became subject to one of the primary objectives of biopolitics, the conservation of population. A vast project to reduce perinatal mortality through a risk-based prenatal care regime was invented during the mid-1950s to the mid-1960s.

My analysis concentrates on the initial part of the perinatal threshold, its prenatal aspects, since I wish to explore the effects of extending the threshold of the living subject into the time of pregnancy and the unborn. The techniques of risk-based prenatal care identified "high risk" pregnancies as the chief target for reducing perinatal deaths.[2] The saturation of pregnancy with risk in our present was designed as a governmental solution to a problem construed as that of reducing "perinatal mortality and morbidity" at the level of population. Under the perinatal threshold of the living subject, pregnancy was construed as a time of risk needing intensified care to prevent death and illness to the unborn from the twenty-eighth week of gestation to the end of the first week after birth.

"Perinatal" is a medical concept. Its emergence is examined here in the context of an international medical research literature. The history of risk governance at the perinatal threshold is approached through a case study – what might be called a microgenealogy – of the Ontario Antenatal Record: a standardized risk assessment device that has been used in the Canadian province of Ontario since 1980.[3] The specificity of the discussion enables a precise description of how prenatal risk assessment was articulated to the reduction of perinatal mortality and entered into clinical practice. The Ontario Antenatal Record comprises a particular example of a text type found across many nations today, including Australia, the UK, the United States and Western European nations. The discussion of prenatal risk assessment returns in the next chapter, which investigates how midwives position themselves in relation to risk governance when routinely completing the Ontario Antenatal Record.

My analysis of the perinatal threshold follows on from David Armstrong's (1986) classic study, "The Invention of Infant Mortality", which examined the history of the subdivision of infant deaths (infant, neonatal and perinatal mortality plus stillbirths) in the official statistics of England and Wales during the first half of the twentieth century. Where Armstrong bases his account on an official source, the *Review and Annual Reports* of the Registrar General of England and Wales, I have privileged published medical research. My choice of source shifts analysis to the history of social medicine and pathological anatomy, and to a period slightly prior to the one Armstrong examines in his discussion of perinatal mortality, as published research tends to precede the stabilization of medical concepts in official statistical regimes. I have added a transitional stage between neonatal and perinatal mortality, "fetal and neonatal" mortality, which appeared as a category in medical research publications, but not in official statistics. The concentration on medical research frames the study of the perinatal threshold in an international[4] rather than the national frame, a study of a transnational discourse. Last, my concern with the formation of risk techniques as a means of reducing perinatal mortality goes beyond the analysis found in Armstrong's article, which ends with the inclusion of perinatal mortality in the vital statistical regime of England and Wales during the mid-1950s.

The perinatal threshold unsettled the birth threshold in the name of optimizing population at the entry to human life. The analysis of risk-based prenatal care provided here is pertinent both to foucauldian scholarship on risk governance and to feminist scholarship on the fetus. With respect to the history of the present, this account shows the invention within the health complex of a risk technique beyond insurance: clinical risk, which involves screening populations and affixing risk classifications to the bodies of persons. Clinical risk marks an adaptation of the calculative, future-oriented and preventive risk schema to the health complex, binding together cheek by jowl disparate forms of medical judgement – the judgement of the normal and the pathological (diagnostics) and the judgement of risk. Heterogeneous forms of medical reasoning are brought together in clinical risk, which optimizes health at the perinatal threshold under conditions of uncertainty.

With respect to feminist scholarship, which has focused on analysing the cultural consolidation of the fetus in diagnostic technologies such as ultrasound (Mitchell 2001; Morgan 1996) and amniocentesis (Rapp 1999), the therapeutic technique of fetal surgery (Casper 1998), popular culture (Hartouni 1997: 51–67; Petchesky 1987; Taylor 1992), and consumption (Taylor 2000), I here show another surface of emergence for the unborn as fetus: routine clinical risk assessment. The use of prenatal risk techniques in order to reduce the perinatal mortality rate attached pregnancy to a comprehensive regime of clinical assessment for the sake of *fetal* health. The proliferation of risk factors around pregnancy was not initially accomplished through the subjectification of the unborn as fetus, but through a biopolitics of population that sought to lower the rate of death and illness during the perinatal interval through the care of the fetus during pregnancy and birth.

The birth threshold and infant mortality

Birth marked the threshold of the living subject among the cultures of Western Europe and the European diaspora at the point at which this investigation begins, the early twentieth century. The conferring of personhood and individuality on the living subject occurred only at the end of the birth process, although its arrival was often anticipated prior to labour. However, the signs of pregnancy were ambiguous, compatible with sickness, growths of various kinds, and impedance of the body's flows (Duden 1993: 83–88; Duden 1991: 123–135). Before hormonal testing became clinically available in the 1930s, it was considered difficult to know whether changes in a woman's body definitively indicated pregnancy. For centuries the condition of being pregnant was first known by women from the bodily experience of "quickening", a feeling of movement within their bodies indicating a new kind of life (Duden 1993: 79–82; Duden

1992). The movements, some of which could be felt by others, including midwives and physicians, were interpreted to mean a woman was "quick with child", that is, expecting a child (McClive 2002). While quickening was treated as a rough proof of pregnancy for all practical purposes, definitive knowledge that a particular woman was pregnant was hedged with interpretive uncertainty. The errors of expertise in confirming pregnancy were notorious. At law, stays of execution for capital crimes were in principle given to women quick with child, but the confirmation of quickening required the assistance of expertise (surgeons and midwives) or, in England, a "jury of matrons" (women who had given birth), and both groups made serious errors of judgement, the stuff of popular song (Ackerknecht 1976; Oldham 1985). So too, the feeling of quickening might cease, leading to uncertainty about whether the woman had been pregnant or whether the pregnancy was no longer quick/viable. Fetal demise was considered difficult to diagnose by physicians into the mid-twentieth century.[5] Under these conditions of interpretive uncertainty, the threshold of the living subject was placed at birth, the zero point of its entry into social time.

Barbara Duden, whose writings have constituted the study of women's experience of pregnancy as an historical object, argues that, during the seventeenth and eighteenth centuries in Europe, birth manifested the "truth" of pregnancy, clarifying whether a woman bore "true" or "false fruit", that is, a child or inanimate flesh such as a mole or a stillbirth (Duden 2002a: 63). Women sensed their "fruit" as a presence that was hidden, secret, not yet revealed – an immanent/imminent presence that Duden terms "*non-dum*": not yet (Duden 2002a: 58–60). Duden suggests that, when a woman felt movements in her womb, the *non-dum* was perceived to be alive, but, in the absence of quickening, the *non-dum* was interpreted as potentially a lump of flesh. The status of the unborn never entered certainty except in the retrospect of birth. During pregnancy, women perceived and conceived of their bodily interiors as a hidden, inner, beyond (Duden 2002a: 60–61; 2000: 32–34), the nature of which was revealed at the crossing over from the interior to the exterior at the threshold of birth.

In an arresting image, Duden observes that practising physicians of the early modern period believed that the unborn were as mysterious as the constellations (Duden 2002b). Visual representations of the unborn customarily imaged male babies, but the images were not understood as realistic representations of the *non-dum,* but rather as emblems that symbolized the unborn (Duden 2002b: 18–30) under conditions where there was no biological concept of a fetus and hence no concept of fetal development. Until the early nineteenth century, practising physicians realized, if reluctantly, that their knowledge of a woman's pregnancy was dependent on her speech, her own accounts of her bodily perception/sensation, with the physical signs (*signa*) that a physician might observe being insufficient for a diagnosis of pregnancy (Duden 2002b: 41–44). At the level of medical practice, physicians recognized that their knowledge of pregnancy was necessarily derivative of and dependent on women's perceptions and accounts of their bodies. In both everyday life and medical care what Duden (2002b: 18; 2002a: 62) terms "women's somatic experiences" (*somatische Wahrnehmungen der Frauen*) was considered the primary reliable evidence about the "coming child".

As Duden emphasizes, the early modern conceptualization and visualization of the unborn began to come apart during the latter half of the eighteenth century in the publications of Haller and Soemmerring (Duden 2002b: 36–41; 1999: 18–24), the result of a epistemological event in medicine: the reading of the unborn as "fetus", that is, a biological, developmental entity undergoing sequential changes in a linearized time. The significance of the biological fetus for medicine, law and female bodily experience during the nineteenth and early twentieth centuries is difficult to estimate at the present time, as the historical literature on the subject is still inchoate. But it has been established that medical research into the fetal heartbeat (first recorded as audible in 1818) took place at the *Maternité* hospital in Paris from the late 1820s and resulted in recommendations for new treatment methods (Herschkorn-Barnu 2002). As well, Nicholas Hopwood (2002 and 2000) has analysed nineteenth-century anatomical drawings of fetal development. The fate of quickening in the nineteenth century remains a topic in search of an author, but one can at least surmise that, by the 1830s in England, both medical jurisprudence and statute law had undermined quickening as the first moment of life (McLaren 1984: 142–3). During this same decade English courts no longer accepted evidence about quickening provided by juries of matrons, attacking the salience of quickening as a concept and rendering obsolete the jury of matrons, the historical judges of quickening in capital cases involving a plea for a stay of execution on this ground (Gavigan 1984: 34–36).

One could multiply examples pertaining to the reconfiguration of the unborn as biological fetus in nineteenth-century medicine, but it would initially appear that the biological fetus was associated with research rather than clinical transformations in the nineteenth and early twentieth centuries, with the possible exception of some elite and destitute women, the former in private medical consultations and the latter in lying-in hospitals. Prior to the twentieth century only these women received medical or midwifery attention during pregnancy. The vast majority of women in Western Europe and North America received no antenatal care until the late 1920s and thus remained remote from the reconceptualization of the unborn as fetus. Marjory Tew, for instance, estimates the numbers of British women receiving antenatal care in 1915 as "inestimably small", rising sharply to 85 per cent in 1935, and then to 99 per cent in 1946 (Tew 1990: 79). In the Canadian case, Mitchinson (2002: 106) holds that, while "[t]he need for prenatal care became an unquestioning belief in the interwar period", provincial rates differed; in 1947, 39.4 per cent of pregnant women in Ontario received prenatal care, with the number rising to 66.5 per cent in 1953. The presence of the fetus was weakly institutionalized even in routine prenatal care until risk-based prenatal care for the purpose of reducing perinatal mortality began to colonize pregnancy during the 1960s. So it was that birth remained the threshold of the living subject until roughly the mid-twentieth century.

The primary objective of prenatal care during the 1920s and 1930s was the reduction of maternal mortality (Oakley 1984: 86), particularly mortality related to "difficult labour" and potentially treatable conditions during pregnancy such as eclampsia (high blood pressure) (Silverton 1993: 102–104; Mitchinson 2002: 104–157; Thompson, Walsh and Merkatz 1990: 14–21). Its objective was not the health of the unborn. In 1929 the British Ministry of Health issued a Memorandum, "Ante-natal Clinics: Their Conduct and Scope," which Oakley (1984: 79) calls the "Magna Carta" of prenatal care because it articulated a standard for British prenatal care that held late into the twentieth century. Whether the 1929 Memorandum directly inspired international commonalities in the standard of prenatal care, or whether national practices arose independently is presently unknown – and there is doubtless a complex history not yet written – but the Memorandum outlines the prenatal care found in mid-twentieth-century urban areas, including Canada. The Memorandum laid out a series of objectives for prenatal care: the prediction of "difficult labour", maternal education and co-ordinated care, and the diagnosis and treatment of toxemia (hypertension), infections, and sexually transmitted diseases (Oakley 1984: 79). It recommended that prenatal care begin at the sixteenth week of pregnancy, when the medical and obstetrical history would be taken, a physical examination performed, and questions pertaining to personal and domestic hygiene asked. The health care provider was to measure uterine height and circumference, test urine, take blood pressure, listen to the fetal heart, and revise the expected birth date (known as EDD, expected date of delivery). As the pregnancy advanced, clinical care was expected to become more frequent: every two weeks between the sixteenth and the thirty-sixth weeks, afterwards weekly. The standards outlined in the Memorandum were recommended without a clear basis in clinical or research evidence, and remain contested into the twenty-first century (Tew 1990).

Twentieth-century population power precipitated an unsettling of the birth threshold as the result of attempts to reduce mortality in the first year of life. The perinatal threshold was forged after various national infant welfare movements had successfully initiated a lowering of infant mortality rates. Mortality statistics comprise a technique for rending bare life as an object of governance. The success of efforts to conserve population in the first year was judged through the technique of mortality rates. In biopolitics mortality rates have, however, represented much more than prosaic counts of the living and the dead. During the late nineteenth and early twentieth centuries, infant mortality rates were mobilized as collective representations of the nation: symbols of solidarity. To this day, the effectiveness of a national health regime is judged by its maternal and infant mortality rates.

Since the first international comparative tables for maternal and infant mortality rates were published in 1913, the infant mortality rate has comprised a measure of national self-worth (Wertz and Wertz 1989: 271). Infant welfare movements in Europe and North America at the turn of the twentieth century dedicated themselves to lowering infant mortality in the name of fighting perils thought to endanger their respective nations. Infant mortality mirrored national fates. Anxieties about population varied both within and between nations, ranging from fears of new immigrants from Southern Europe displacing the descendants of those from Northern Europe (Canada and the US), concerns about the lower birth rates of the bourgeoisie and middle class as compared with the working class (Britain, Canada, France, US), worries about an association of falling birth rates with national decline (France and US), and dread of imperial decay coupled with fears about the well-being of the "national stock" in light of the poor health of military recruits and deaths in warfare (Britain, Germany).[6] The genealogy of infant mortality statistics opens to a running theme in twentieth-century population power: the ranking of populations within and across nations.

A number of measures were offered as solutions to the international demographic anxieties that continued after the decline in the infant mortality rate: public education in infant care, maternalist legislation and policy, and prenatal care. These were intended to improve the quantity and quality of population through the reduction of both maternal and infant mortality. All three measures predated World War I, but spread rapidly during that war, and were consolidated during the interwar period, with national variations in the time, degree and form of implementation. Prenatal education programmes trained women in nutrition and hygiene during pregnancy as well as in infant care (Arnup 1994: 32–83; Comacchio 1993: 92–115). In Western Europe and North America, the period from the late nineteenth century to the early twentieth saw the beginnings of maternalist social welfare, including maternal protection laws limiting or prohibiting women's participation in the paid labour force, paid maternity leaves from work, and family allowances.[7] These legislative initiatives and policies attempted to provide incentives for having children under conditions where states feared "race suicide" as regimes of vital statistics produced evidentiary knowledge showing that various national populations were falling below replacement rates (Bock and Thane 1991; Klaus 1993b). Maternalist politics was broadly pronatalist, but also aimed to reduce maternal mortality, concern for which dominated the interwar years, succeeding the international emphasis on mortality during the first year of life that had dominated the prewar period (Comacchio 1993: 64–91). In the US, federal policies targeted infant mortality under the terms of the federal Shepphard-Towner Infancy and Maternity Protection Act (1921), which provided subsidies to states establishing prenatal and postnatal clinics for mothers; Shepphard-Towner was a response to statistical data showing the American infant mortality rate to be higher than that of other Northern nations (Skocpol 1992: 10 and 490–494). A generation of feminist historians has documented the formation of maternalist welfare states in all their national specificities; the only and rather bald point I wish to make here is that the maternalist welfare state was in part a response to the problematization of infant mortality across national divisions.

By the time of World War I, physicians, infant welfare organizations and departments of state had called attention to deaths in the first month of life. In France the physician Pinard – a nationally known figure whose work on infant mortality postdated that of Budin and Tarnier – had as early as the mid-1890s recommended a programme for the reduction of early infant mortality due to prematurity,[8] what he called *puériculture*: maternal education, maternity leave and municipal shelters for working-class women (Baker 2000: 323–324). In Scotland John W. Ballantyne[9] likewise proposed prenatal care for the reduction of "neonatal mortality" in his much read *Antenatal Pathology and Hygiene*, Vol. 1: *The Foetus* (Ballantyne 1902: 1, Figure 2). While Ballantyne gestured to the importance of lowering neonatal mortality because of the falling birth rate in England and Wales, his arguments were not primarily statistical, focusing instead on the need to provide pregnant women who were ill and/or destitute with rest and care in maternity wards (Ballantyne 1901).

During the late nineteenth century and the first decade of the twentieth, a scattering of international reports and articles had remarked that infant mortality was highest in the period after birth. The analytic limitations of the infant mortality rate were questioned in England as early as 1905 by Sykes, the Medical Officer of Health for St Pancras (a district in London, England). Sykes showed that infant mortality for the year 1905 in St Pancras was greater in the first three months of life than for any of the succeeding quartiles (McCleary 1933: 32–343). Beginning in 1906 the *Annual Reports* of the Registrar-General for England and Wales gave infant mortality not only by year, but for each of the first four weeks, and by month (Newsholme 1910: 979; Tew 1990: 232).

During the last years of the nineteenth century in the US it became common practice to subdivide infant mortality into two categories: deaths in the first month and deaths during the rest of the first year (Brosco 1999: 481). At a federal level, the 1913 Census Report (a national compilation for the year 1910) divided time of death by first day, first week, first month and first year. The 1913 Census Report showed that infant mortality had fallen off after one month, but not before (Meckel 1990: 168–169). The American infant welfare movement had reached the same conclusion prior to the release of the 1913 Census Report, causing it to rethink its tactics from its prior advocacy for purifying the milk supply in urban areas, to a new plan for educating mothers, hoping to improve nutrition and hygiene during pregnancy as well as infant care (Meckel 1990: 166).

In tandem with mortality statistics, neonatal mortality was problematized by pathological anatomy during the 1910s and 1920s. Recommendations for the reduction of neonatal mortality were made on the basis of postmortem examinations and autopsy studies that sought to establish cause of death among newborns. The autopsy studies consistently reported high levels of preventable mortality from "complications of labour", singling out the dangers of high forceps deliveries as a cause of both neonatal deaths and stillbirths (Holland 1922; Holt and Babbitt 1915; Palmer 1928; Serbin 1928).

In the Euroamerican states considered here, the mortality rate among newborns was recognized into the 1930s as significantly higher than the mortality rate for the remainder of the first year (Bonar 1932, 1934, 1935; Duncan 1937; Eden 1931; Grulee 1933; Harer 1932; League of Nations Health Organisation 1930). Trying to account for differences in the mortality rate during the first year, American pediatrics during the 1930s concertedly investigated the causes of neonatal mortality and stillbirths. A series of publications during that decade proposed "prematurity" as the leading cause of neonatal mortality (Bonar 1935; Bundesen *et al.* 1938; Clifford 1936; Lyon and Bemis 1931; Murphy and Bowman 1932),[10] which became a consensus position in US and Canadian medicine by the latter half of the 1930s.

Unlike neonatal deaths, where death occurred after live birth, stillbirths were excluded from the computation of the infant mortality rate. An international consensus existed with respect to the definition of infant mortality in the late nineteenth and early twentieth centuries; infant mortality consisted of deaths in the period from birth to the end of the first year (Brosco 1999: 478). Only live births – not stillbirths – were included in comparing the infant mortality rate. The framing of stillbirths within medicine and official discourse as preventable deaths that arose from "natural causes" marked a sharp break from their prior association with infanticide and, in nineteenth-century cities, abortion (Baker 1996: 22; Mooney 1994: 53; Ward 1981: 42–44).

Early studies problematizing cause of death among stillbirths began appearing in the mid- to late 1910s (Holt and Babbitt 1915; McQuarrie 1919; Routh 1913; Williams 1915). In order to establish cause of death, these early studies of stillbirths combined clinical information such as fetal presentation (position in uterus) and type of delivery (spontaneous, forceps, etc.) with postmortem findings about lesions. Data included cases of both stillbirths/fetal deaths (after viability) and neonatal deaths. Holt and Babbitt (1915) disaggregated the data for stillbirths, enabling them to distinguish cause of death in stillbirths from that in newborns. They listed the following as causes of "fetal death" (in descending order of numerical significance): prolonged, difficult or complicated labour, toxemia of pregnancy, syphilis, prematurity, "monsters", unknown (Holt and Babbitt 1915: 290). McQuarrie's investigation at the University of California Hospital combined the analysis of stillbirths with deaths in the first twelve hours to produce the following list of causes of "fetal deaths": syphilis, unknown, birth trauma, toxemia, fetal abnormality, prematurity, placenta praevia and various (McQuarrie 1919: Table 1: 1574).

During the 1920s, stillbirths were first problematized as preventable, a transformation linked with medical research showing that pregnant women diagnosed with syphilis who received treatment were much less likely to have stillbirths and to transmit the infection to their children *in utero*/in labour than those left undiagnosed and untreated (Ballantyne 1922: 587–588; Cruickshank 1922; Detweiler 1926: 545–546; MacMurchy 1926: 442–443; "Maternal Syphilis and Fetal Mortality" 1920). The autopsy reports of the 1920s and early 1930s provided a gruesome litany of unnecessary stillbirths and neonatal deaths caused by poor obstetrical management of childbirth; this was the heyday of high forceps delivery. One indication of the changing status of stillbirths is provided in the recollections of John Ballantyne at the annual meeting of the British Medical Association in 1922. Ballantyne (1922) compared the current session on the theme of "stillbirths and neonatal deaths" with the session on "intrauterine death" that he had attended in 1888, noting a number of differences. He recalled that, in the 1888 discussion, "the idea of supervising all pregnant women in the interest of the few who were in danger of giving birth to dead infants was absent"; syphilis was untreatable; and no one discussed compiling statistics on stillbirths, fetal deaths from and during labour, or neonatal deaths from causes arising during gestation (Ballantyne 1922: 583). By 1922 a reduction in the numbers of stillbirths had become imaginable.

The gaze of pathological anatomy constituted stillbirth as mortality – knowable and preventable mortality. During the 1920s postmortem and autopsy studies overwhelmingly linked fetal death with poor obstetrical methods during childbirth. Holland (1922), a surgeon at the City of London Maternity Hospital, performed postmortem examinations on 300 stillbirths between 1914 and 1921. His report identified the four leading causes of death as follows: 51 per cent due to complications of labour, 16 per cent syphilis, 10 per cent maternal toxemia, and 2 per cent renal and other maternal diseases (Holland 1922: 125). He estimated that 52 per cent of stillbirths were preventable through improved prenatal care and obstetrical methods (Holland 1922: 127). The later works of Holland and Lane-Claypon (1926), Cruickshank (1930), and Palmer (1928) were exceptional at that time for using autopsies to investigate cause of death in infants and stillborns.

Coinciding with the studies of pathologists that problematized stillbirths as preventable mortality, vital statistical regimes constituted stillbirths as a subpopulation whose mortality rate might be decreased. While counting stillbirths had become common during the period of the infant welfare movement, the reporting of fetal death was not made compulsory for the purposes of many Euroamerican vital statistical regimes until the late 1920s, with the exception of New Zealand, where stillbirths were made notifiable as early as 1913 (Sutherland 1949: 2–5). In 1925 the League of Nations Health Organisation first adopted a definition of stillbirth as: "the birth of a (viable) fetus after at least 28 weeks pregnancy in which pulmonary respiration does not occur. Such foetus may die either a) before, b) during, or c) after birth, but before it has breathed" (quoted in *Causes of Stillbirth, 1943–55*, 1957: Preface [n.p.]). The League of Nations' definition was consistent both with prior nineteenth- and early twentieth-century medical and legal understandings that the fetus was not viable until the seventh month of pregnancy; in nineteenth-century law: "no Western country regarded a fetus below 180 days as viable" (Baker 1996: 22). The definition of stillbirths endorsed by the League of Nations Health Organisation was widely used after it became a category of report for the member nations of the League; by way of example, the League's definition was adopted by the national statistical systems of Canada, England and Wales, and the United States during the latter half of the 1920s.[11]

The reduction of stillbirths had fallen outside the programme of biopolitics found in infant welfare movements of the late nineteenth and early twentieth centuries. Where infant welfare movements had targeted deaths throughout the first year as preventable, during the interwar years research in medical statistics and pathological anatomy problematized a narrower interval: the deaths of newborns and stillbirths. The constitution of stillbirths as preventable would require techniques capable of preserving fetal life late in pregnancy and childbirth – action at the threshold of the living subject. The techniques for decreasing neonatal and infant mortality recommended in the research literature of the 1920s sketched the beginning of a programme for decreasing neonatal and stillbirth rates: improved obstetrical care during childbirth, anti-syphilis testing and treatment, specialized care units in hospitals, and educational programmes during pregnancy.

Fetal and neonatal mortality

Fetal and neonatal mortality became a single mortality in the medical research publications of the 1940s: a reconceptualization of the threshold of the living subject as encompassing the period before and after birth. The association of fetal with neonatal began to appear in medical research of the 1930s and 1940s, and was consolidated in the late 1930s and early 1940s under the dual impact of pathological anatomy and the statistics of social medicine: babies born dead and infants dying shortly after birth were coming to resemble each other. The phrase "fetal and neonatal" was used to suggest physiological and pathological continuities of bodily substance across birth. In his study of English and Welsh vital statistics, Armstrong (1986: 215) observed this change: "[W]hereas in the nineteenth century stillbirths were joined to miscarriage and abortion as manifestation of foetal loss, in the inter-war years they become the natural partner of neonatal mortality, separated only by a single breath." Newborns were analytically disconnected from the remainder of life in the first year, and attached to the unborn configured as fetus through an argument that minimized birth as a physiological boundary. "Fetal and neonatal mortality" had by the 1940s emerged as a standard phrase in medical publications: two sides of a common mortality linked in ways that were simultaneously self-evident and obscure.

A small number of research studies during the 1920s of fetal with neonatal mortality reflected increasing medical concern with the analysis of cause of death in still births and the prevention of death among newborns (Browne 1921; Thomson 1920; Willson 1924). In 1923 the obstetrician William Cosbie published an article in the *Canadian Medical Association Journal* that attracted interest.[12] He argued that fetal and early infant mortality were both preventable, but had been neglected because "the problems of causation and prevention of maternal morbidity and mortality so overshadow the question of the loss of foetal and infant life" (Cosbie 1923: 877). Cosbie combined stillbirths "after the period of viability had been reached" with mortality in the first two weeks to form a measure of "stillbirth and early infant mortality" per 1,000 births (ibid.). (Cosbie did not explicitly state whether the denominator combined live births with stillbirths.) The joining together of stillbirths with mortality in the first two weeks gave Cosbie's argument rhetorical force by creating a single mortality rate that could be counterposed to maternal morbidity and mortality. He divided causation into four groups: deaths with no assignable cause, deaths due to antenatal causes, deaths during birth, and postpartum deaths where pregnancy and birth had been "normal" (Cosbie 1923: 877–880); the first two were common to stillbirths and newborns, the third pertinent only to stillbirths and the fourth to newborns. His recommendations for preventing "stillbirth and early infant mortality" mainly involved improved obstetrical care during birth, and regular prenatal care leading to the diagnosis and treatment of syphilis and eclampsia.

In Austria, the pediatrician Sigismund Peller,[13] who had trained in social medicine with Ludwig Teleky[14] at the University of Vienna (Peller 1967: vii–viii), published a long series of articles from 1923 to 1940 on stillbirths and infant mortality. His work during the 1920s and 1930s is significant because it precedes his English-language articles on "perinatal mortality" written in the 1940s, which introduced the term into English from its German-language uses of the late 1930s. Peller undertook the quantitative analysis of stillbirth and infant mortality, with a particular interest in mortality during early infancy, examining how mortality varied with respect to such factors as nutrition, marital status, birth order and clinical care during pregnancy. Peller's (1923a and 1923b) first publications on stillbirths and infant mortality appeared in the same year as William Cosbie's article, 1923.[15] Like Cosbie, Peller was concerned with the relation between stillbirths and early infant mortality, separating deaths in the first year into a series of smaller intervals, beginning with deaths in the first week. These 1923 studies demonstrated that the rate of stillbirths and deaths in the first week depended on living conditions late in pregnancy. Peller showed that women who had stopped domestic and paid work late in pregnancy through voluntary placements in maternity homes had reduced rates of stillbirth and early infant mortality compared to those who remained at work. Although Peller included stillbirths and early infant deaths in his data analysis, his 1923 articles differed from Cosbie's in not creating a combined measure.

By the early 1930s Peller had begun performing joint counts of stillbirths and neonatal deaths (Peller 1930, 1931b, 1931c), consistently arguing that the rate of stillbirths and neonatal mortality depended on a range of social factors late in pregnancy. On this basis Peller believed that preventive practices after birth would be of little influence in reducing fetal and neonatal death (Peller 1930: 848). He instead advocated what he regarded as more effective labour protection and social welfare measures that would act on the unborn during pregnancy, specifically the prohibition of paid work in the last four weeks of pregnancy with state replacement of lost income, social insurance for domestic help during the last six weeks of pregnancy, and the building of maternity residences for women whose homes were "unhygienic" (ibid.). Peller's research problematic involved the relation between social class, stillbirths and neonatal mortality, a problematic that led to his privileging of what he called "exogenous factors" (Peller and Bass 1924) (e.g. nutrition, rest) as determinants of fetal development in opposition to hereditarian explanations and assumptions that the fetus developed autonomously, if necessary stripping the maternal body for its nutritional needs. His programme for reducing fetal and neonatal mortality rates was consonant with legislation for maternal protection and maternal benefits in the building of welfare states in Western Europe and North America during the interwar years.

As discussed in the previous section, from the period of the mid-1910s British and American medical research had linked stillbirths and early infant deaths etiologically through common causes of death. These early classifications of cause of death (Holt and Babbitt 1915; McQuarrie 1919) for the most part attributed death to maternal disease or complications during birth. Later research used autopsy findings to develop classifications of cause based on fetal pathology or combinations of clinical and pathological findings (Browne 1922; Palmer 1928; Serbin 1928).

Two key researchers systematized the study of "fetal and neonatal mortality" from the mid-1930s to the mid-1950s, one, Edith Potter,[16] an American specialist in pathological anatomy, the other, Dugald Baird,[17] a Scottish obstetrician. They consolidated stillbirths and neonatal deaths as fundamentally similar in their social and pathological causes of death. In Potter's words (Hoffman 1982: 1552), prior to the first edition of her book (with Fred Adair), *Fetal and Neonatal Death* (1940),[18] "[W]e talked about infant deaths and stillbirths, but we had not yet recognized it as a special field with special problems."

Potter's work[19] on fetal and neonatal pathology began in 1935, the same year that the American Academy of Pediatrics had called for research and action on neonatal mortality (Round Table Discussion on Prematurity 1936). Fred Adair[20] had hired Potter at the Chicago Lying-in Hospital to do research on the causes of fetal and infant mortality. He formed an alliance with Herman Bundeson, Commissioner in the Chicago Health Department, who believed that infant mortality was the best indicator of urban health. Bundeson had the ambition of making Chicago the healthiest city internationally, a common if appalling ambition in the agonistic field of comparative mortality statistics; truth regimes speak with monotonous regularity. Convinced by Adair's argument that it was necessary to know the cause of infant deaths in order to prevent infant mortality, Bundeson levered the fact that, since his department provided burials free of charge for stillbirths and babies dying in the first month, the performance of an autopsy could be required as a condition for issuing a burial permit (Hoffman 1982: 1552). This practice continued a long history of using the bodies of the poor and destitute for anatomical specimens (Richardson 1987).

Where Peller had linked fetus and newborn on the basis of quantitative data associating mortality and survival with the "life conditions" of women during pregnancy such as nutrition, work and marital status, Potter as a pathologist took as her object a three-dimensional corporeal space, seeing in the organs found there the marks of lesions revealing cause of death. Potter's first article on fetal and neonatal mortality, co-published with Adair in 1936, constructed the threshold of the living subject as a continuous interval before, during and two weeks after birth. Adair and Potter observed that mortality during the first year had fallen over the previous twenty years, but little change had occurred under two weeks, and none for stillbirths. To a pathologist the pattern suggested the possibility of differing causes of death, one group of causes after two weeks and a second for fetuses and newborns. During the first two weeks of life "[t]he causes of death ... are largely distinct from those occurring late in life, but are closely allied with those operating as a cause of death before and during birth" (Adair and Potter 1936: 281). Their article divided cause of death into three periods: antepartum, intrapartum and postpartum. If the fetus and neonate died of similar causes before, during and after birth, then the stillbirth rate might potentially be reduced simultaneously with the neonatal mortality rate: "Attempts to lower the neonatal death rate cannot be made without also attempting to lower the stillbirth rate, for deaths in both groups are frequently due to the same fundamental causes" (Adair and Potter 1936: 286).

For her 1940 book with Adair, Potter was to perform approximately 4,000 autopsies on fetal deaths over twenty weeks' gestation, correlating autopsy findings with maternal and infant histories to determine probable cause of death, definitively establishing that "the causes of fetal death are in general similar to those responsible for loss of life during the first few days after birth, so that the same problems presided in each group" (Potter and Adair 1940: xv). Potter's early work in pathological anatomy showed that intracranial hemorrhage and anoxias (lack of oxygen) from traumatic and long labours were the leading causes of fetal and neonatal mortality (Potter and Adair 1939: 1550; Potter and Adair 1943: 1065).[21] Pathological anatomy thus opened to a series of recommendations to prevent fetal and neonatal mortality, pointing to better obstetrical care through "the avoidance of dangerous procedures", the use of medications to prevent premature birth, and "proper methods of resuscitation" (Bundesen *et al.* 1938: 141).

The concept, "cause of death", acted to bring together newborns and fetuses across the boundary of birth, making them fundamentally similar with respect to pathology and unlike later deaths. Fetuses and newborns were seen to suffer the same pathologies, Potter and her colleagues argued, and thus their mortalities were as one (see Figure 2.1). The aim of reducing mortality after birth had led to the elision of birth as the lower and absolute limit of life for the purposes of computing mortality.

Where Potter's research fell strictly within pathological anatomy, combining autopsy results with clinical information, Baird's research used the survey

TABLE 6.—*Final Diagnosis of Cause of Death of 2,000 Fetuses and Infants Based on Clinical and Anatomic Evidence*

	Number	Per Cent
Prematurity	416	20.7
Asphyxia	324	16.1
Birth trauma	255	12.7
Malformations	209	10.4
Pneumonia	132	6.5
Erythroblastosis	39	1.8
Syphilis	26	1.2
Miscellaneous	22	1.1
Maternal toxemia without other cause	97	4.7
Unknown		
1. Premature	231	12.5
2. Term	249	12.3
Total	2,000	100.0

1. Antepartum deaths before term.
2. Includes 145 antepartum deaths at term.

In this table Potter grouped together fetal deaths in late pregnancy and infant deaths during the first month as part of the same set with equivalent causes of death.

Figure 2.1 Potter on the causes of fetal and neonatal mortality

Source: Potter, Edith (1940) "Fetal and Neonatal Deaths." *JAMA* 115 (12): 998, Table 6.

methods of obstetric epidemiology, part of a British research trend during the period between 1935 and 1950 that investigated social class differentials in health, particularly mortality and fertility differentials (Oakley 1991: 188). Baird's problematic was the effect of social class on the rate of stillbirths and neonatal deaths, using primary research from the Aberdeen Maternity Hospital. Previous researchers had shown that the infant mortality rate varied according to income, being highest among the poor. Baird extended this inquiry into fetal and neonatal mortality, a process that led him, as it had Peller, to question the "autonomy" of the fetus, the commonly held position that the fetus had the capacity to develop independent of external influences. Using the analytic power of the mortality rate to standardize measures and establish fields of comparison (Ewald 1990), Baird made a case that the stillbirth and neonatal mortality rates were dependent on social class, and that these rates might be reduced through social policies promoting better housing and nutrition. Baird worked in Aberdeen, a Scottish city with a long history of working-class and underclass poverty, and his research showed an association between social class and "foetal mortality" (Baird 1947), linking stillbirths and early infant deaths to poverty and malnutrition (Tew 1992: 232).

Baird's 1941 article with John Wyper in *The Lancet* began squarely in concerns about decreases in the national population of England and Wales. Their population projection repeated that of Richard Titmuss, who, in turn, had repeated Enid Charles in stating that these populations would decrease by 24 per cent in each generation if their current fertility rate continued into the future (Baird and Wyper 1941: 657). Baird and Wyper argued that regional variations in infant mortality rates across England, Scotland and Wales proved the existence of preventable mortality, a finding they hoped would assist in increasing their national populations (Baird and Wyper 1941: 657). Reasoning that infant mortality and possibly neonatal mortality were associated with poverty, they went on to argue, in a seeming *non sequitur*, that neonatal mortality should be combined with stillbirths. The combination of stillbirths and neonatal mortality would give insight into those "children" dying after "viability": "Since most neonatal deaths are directly connected with pregnancy or childbirth, it is permissible to add them on to the stillbirths to get some idea of the number of children who reach the stage of viability and yet die as a result of some complication of pregnancy, labour or the puerperium" (ibid.). This analytic move set the basis for an examination of the causes of death in stillbirths and newborns, with stillbirths cast as preventable mortality, a "wastage of child life" that might be conserved through improved economic conditions, together with better nutrition and rest in pregnancy (Baird and Wyper 1941: 659). Thus Baird and Wyper aligned the reduction of the stillbirth and neonatal mortality rates with the conservation of English and Welsh national populations.

In a 1942 session of the Section of Obstetrics and Gynaecology at the meetings of the Royal Society of Medicine, Baird commented on recently published figures showing Scottish rates for stillbirths and neonatal mortality to be considerably higher than the English and Welsh rates.[22] Baird argued that this was due to the fact of greater poverty and poorer housing in Scotland. His leading recommendations were for better nutrition and rest during pregnancy and better training for women in infant care ("Discussion on Stillbirth and Neonatal Mortality" (1942–1943): 60). When another speaker in the same session, Dr Charles McNeil, explicitly argued that stillbirths and neonatal deaths ought to be treated as a single mortality rate, Baird here as elsewhere took this as given.

The work of Baird and the MRC Obstetric Medicine Research Unit at Aberdeen was accepted in Britain during the 1950s as definitively establishing the significance of "social factors" as a cause of fetal and neonatal mortality. Between 1945 and 1955, Baird and his colleagues demonstrated the positive correlation of poverty, increased age, shortness of height, and "high parity" (number of previous pregnancies) with fetal and neonatal deaths and illness (Baird 1945; Baird 1947; Duncan, Baird and Thomson 1952; Baird, Thomson and Duncan 1953; Baird and Thomson 1954). Baird's work during the 1950s and 1960s combined a multicausal factor analysis incorporating social, pathological and clinical factors to explain what he and his co-authors called "obstetric death" (stillbirths and first week deaths) in a joint 1954 article (Baird, Walker and

Thomson 1954). *The Lancet* was later to credit Baird's work in Aberdeen with setting the framework for the later, national surveys of perinatal mortality in England and Wales (*Lancet* 1980: 840).

The period from 1930 to 1950 saw the consolidation of "fetal and neonatal mortality" as a concept from two directions: social medicine and pathological anatomy. The social medicine of Peller and Baird continued themes that had previously been found in the vital statistics of the infant mortality movement: the association of mortality with social class, poverty and the city (e.g. Newsholme 1910). However, unlike the earlier analyses of infant mortality, which focused on deaths after live births, the social medicine of the interwar and postwar period also analysed fetal death/stillbirths as a form of preventable mortality, linking those born dead and those born alive but dying soon after birth as socially identical. The gaze of the pathologist saw common causes of death in the depths of fetal and neonatal bodies, using the combined findings of autopsies, biochemical tests, and clinical histories to construct causation. Pathological anatomy bonded together "fetus and neonate" as similar in bodily substance due to the fact they shared the same causes of death. Jointly and separately, social medicine and pathological anatomy constituted the fetal and neonatal interval as continuous, unified etiologically across pathological, clinical and social causes of death.

"Fetal and neonatal" became a common phrase within medicine of the late 1940s, suggesting physiological and pathological continuities of bodily substance across birth. Thus, for instance, Clement Smith, in his textbook for pediatricians that was standard in North American medical education during the late 1940s and 1950s (Avery 1992), *The Physiology of the Newborn Infant* (1946), justified the amount of space he had devoted to the fetus, noting that his book was in many ways – despite its title – a study of "fetal and neonatal life". Smith argued that the study of prematurity necessarily implicates the fetal period:

> it is necessary to begin somewhere near the beginning, which is certainly not the moment of parturition. The threads of biological continuity cannot be taken up at unattached and unrelated ends (which, indeed, they do not naturally possess), without regard to the living fabric of integration in which some of them fulfilled as essential a function before birth as after they became rewoven into another pattern. It is also necessary to understand what we can of the difficulties confronting prematurely born infants, and this obviously calls for some grounding in the development of fetal existence.
>
> (Smith 1946: 3–4)

The "fetal and neonatal" interval moved the threshold of the living subject prior to birth, making the threshold coincide with the last weeks of pregnancy.

From the preceding it is evident that the invention of "fetal and neonatal" mortality was not in any way motivated by a political project to claim fetal personhood, nor was it organized around visual representations of the fetus.

"Fetal and neonatal" mortality was constituted in population power: the reduction of infant mortality pursued with such intensity that it crossed over into the birth process and pregnancy. However, it was only during the 1950s when the concept of "fetal and neonatal" mortality came to be revised as "perinatal" mortality and attached to the techniques of risk-based prenatal care that the birth threshold of the living subject was unsettled.

The perinatal interval

The concept of "fetal and neonatal" mortality was renamed "perinatal"mortality during the 1950s, a neologism introduced into English by Sigismund Peller, who, fleeing Vienna in 1934 during the rise of European fascism, arrived in the United States to begin again as a pediatrician. "Perinatal" named a peak on a graph that ambitiously measured mortality from conception to the end of the first year after birth.[23] The term and the graph were the work of Meinhard von Pfaundler (1872–1947), an Austrian pediatrician, Professor at the University of Munich, and Director of the Children's Health Clinic.[24] Pfaundler was an eminent physician known for his publications on the diseases of children,[25] physiology and nutrition in newborns (Pfaundler 1924), prematurity, and miscarriage. "Perinatal" appeared in Pfaundler's 1936 report, "Studien über Frühtod, Geschlechtsverhältnis und Selektion", ["Studies on Premature Death, the Sex Ratio and Selection"], issued by the Children's Health Clinic at the University of Munich.

Pfaundler's 1936 study attempted to form an estimate of the "intrauterine mortality rate" together with its sex ratio. He argued that the importance of the "intrauterine mortality rate" lay in its displacement of birth as the beginning point for estimates of mortality. In place of birth Pfaundler substituted a point he regarded as more appropriate: "*der Anfang des individuellen Daseins, der Befruchtungstermin*" ["the beginning of individual being, conception"] (Pfaundler 1936: 202). In the passage where Pfaundler first used "perinatal", he introduced it as a means of linking time before, during and after birth. Commenting on a graph (see Figure 2.2) that had its horizontal axis divided into monthly intervals, beginning at conception and finishing at the end of the first year after birth, Pfaundler stated: "*Kurz vor, während und kurz nach der Geburt, einer Periode die zweckmäßig als die* perinatale *zusammengefaßt und bezeichnet werden kann, bildet die Mortalität in beiden Geschlechtern, aber noch mehr im männlichen als im weiblichen, eine hohe und steile Zacke*" (Pfaundler 1936: 201; his emphasis). ["Shortly prior to, during and after birth, a period which can usefully be summarized as and labelled perinatal, a high and steep spike of mortality is formed for both sexes, although more pronounced for males than females."] Pfaundler's graph plotted mortality as continuous from conception to the end of the first year of life. At the top of the graph appear three temporal divisions: "embryofetal period", "perinatal period" and "later infancy". The "perinatal period" names a spike in mortality running from the middle of the ninth

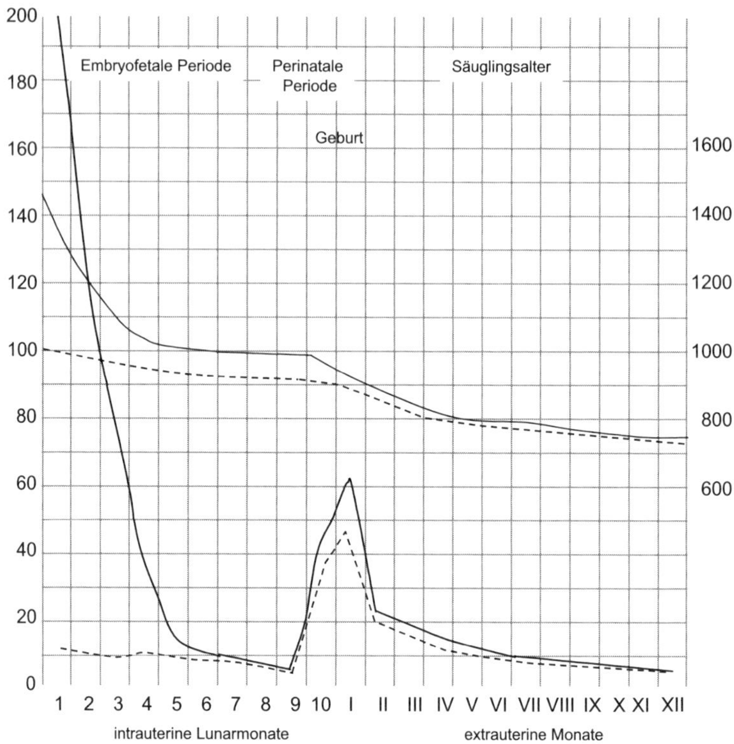

Pfaundler's graph estimates mortality from conception to the end of the first year after birth. The broken line indicates female mortality, and the unbroken male. Following the slope of the lower curves above, one sees a peak in both female and male mortality between the ninth month of pregnancy and the beginning of the second month after birth. Pfaundler called this the "perinatal period", producing a unified temporal interval from what had previously been divided by birth: the time of pregnancy (in common culture) and gestation (medicine) as distinct from the time of babies/newborns/infants.

Figure 2.2 Pfaundler's perinatal period

Source: Pfaundler, Meinhard von (1936) "Studien über Frühtod, Geschlechtsverhältnis und Selektion." *Zeitschrift für Kinderheilkunde* 57: 203.

month of pregnancy, peaking in the first month after birth and declining sharply during the second postnatal month.

The "perinatal" was fashioned as a temporal interval by cobbling together two prior and distinct temporal periods separated by birth: the time of pregnancy and the first year of life. In Figure 2.2, Pfaundler displaced pregnancy with "intrauterine months", that is with the gestational time of fetal development. Figure 2.2 was spatialized in terms of the uterus rather than pregnancy, with two locations in relation to the uterus: inside and outside. The everyday distinction between before and after birth was retained on the horizontal axis of Pfaundler's graph, which contained two numbering sequences: Arabic numerals from 1 to 10 for "intrauterine" months, followed by Roman numerals for the first twelve "extrauterine" months. The temporal axis of a graph conventionally measures time as continuous within the interval chosen, spreading out from the vertical axis in a numerical series. In Pfaundler's graph the zero point where the horizontal and vertical axes met marked the point of conception. The horizontal axis extends from the zero point of conception in two successive series, the first for "intrauterine" months (numbered 1 to 10) and the second for "extrauterine" months (numbered I to XII). "Perinatal" is a medical division of time formed from sticking together two previously heterogeneous, incommensurable temporal intervals on either side of birth: the months of pregnancy and the first year of life. The temporal continuity/discontinuity of the horizontal axis was made possible by, and in turn made possible, a conception of mortality crossing birth.

Pfaundler's neologism used a prefix, "peri", commonly found in medical classification. Within anatomy, "peri" names the area around an organ, tissue or other anatomical point. Medical dictionaries contain dozens of entries with "peri" as a prefix: e.g. the pericardium (the tissues around the heart) and the periosteum (membrane surrounding a bone). The significance of "peri" as a marker does not lie in abolishing a distinction between an anatomical point and what surrounds it, but rather in connecting a point to a delimited surround. By analogy, "perinatal" investigates the time around birth but does not dissolve birth as a temporal marker. Perinatal problematizes the time around birth, creating a temporal continuum from the separate moments of before, during and after. Birth marks an internal distinction within the time of the perinatal, which is divided into the antepartum, intrapartum and postpartum (before, during and after birth).[26] Birth as an absolute threshold holding apart inside and outside is displaced through its integration as a component of a tripartite temporal sequence.

Pfaundler (1936: 202) constructed an image of terrible attrition for the German nation as a consequence of "deaths" in the "intrauterine months": "*Auf eine Minute entfallen in Deutschland etwas mehr als 2 Frühtodfälle*" ["During each minute in Germany a little more than two more premature deaths appear"] (Pfaundler 1936: 202).[27] The total number of deaths from diseases such as tuberculosis and cancer over the entire lifespan, he commented, was just a "bagatelle" compared with premature mortality (Pfaundler 1936: 202).

Sigismund Peller, who like Pfaundler trained as a pediatrician in Austria, claimed the concept of "perinatal mortality" as his own invention, arguing that he had used it consistently in his publications from 1923 onwards, but this had not been recognized (Peller 1967: 158 and 184). Peller appears to have introduced the term "perinatal" into English in the context of a series of articles he wrote during the 1940s on European historical demography (Peller 1943, 1944, 1948). The series of papers demonstrated social class differentials in what he explicitly called "perinatal" mortality for the first time in 1944 and then again in 1948 (Peller 1944: 380 and 1948: 411, n. 1 and 423). He did not employ "perinatal" in his German-language publications of the 1920s and 1930s, although he was one of the researchers who added the stillbirth rate to the neonatal mortality rate to form a combined rate, a measure of fetal and neonatal mortality.

Peller (1979: 148–150) fled Austria in 1934, landing in New York (via Palestine) during 1936, where he fruitlessly attempted to interest American physicians in his work on fetal and neonatal mortality. Trying to convince Cosack, Chair of the New York State Committee on Maternal and Neonatal Death, that fetal and neonatal mortality was primarily social in causation and needed a social planning response, Peller was told that "This is un-American thinking" (Peller 1979: 164–165). Adair (of Potter and Adair) likewise rebuffed him. This was hardly surprising given that Peller had long argued that improved obstetrical methods during childbirth would do much less to decrease fetal and neonatal mortality than social welfare measures. Adair, however, was committed to precisely the opposite position, having hired the pathologist Edith Potter to investigate causes of death among stillbirths and newborns, a project that identified the leading cause of fetal and neonatal mortality as the mismanagement of childbirth by obstetricians (Potter and Adair 1939). Peller's training and commitment to social medicine clashed with Adair's orientation to the study of fetal and neonatal mortality in terms of pathological anatomy: a standard conflict between explanation in terms of pathological vs. social causation. Nor would Peller have benefited from the American anti-Communism that met European refugee intellectuals who were fleeing fascism. Peller received a warmer reception in the later 1940s from Carl Eberhardt (Director of the Bureau of Records and Statistics, New York City Department of Health), who was located in public health. Eberhardt introduced Peller to the organizers of the World Population Conference in 1954, where he served as a discussant in the meeting on "Mortality Trends, with Special Attention to Areas of Lower Mortality" (*Proceedings of the World Population Conference* 1955: 5 and 32).[28] Although Peller introduced the term "perinatal" into English, his contribution was never acknowledged. Until the end of his life Peller remained bitter about what he regarded as his unrecognized work on perinatal mortality.

The question of a connection between eugenics – potentially fascist eugenics – and the formation of the concept of perinatal mortality arises here in the context of German and Austrian history of the 1930s and 1940s. The centrality of demographic and fertility analysis to European fascism (Weindling 1988) makes the question obvious. Perhaps surprisingly, the association between eugenics and the concept, "fetal and neonatal mortality"/"perinatal mortality" appears weak. It should initially be borne in mind that, at least in the period before and during World War I, eugenicists were dubious of calls to lower the rate of stillbirths and neonatal deaths, as they regarded these deaths as a natural mechanism for weeding out "weaklings" and the "unfit" afflicted with "hereditary taint" (Newsholme 1910: 9; Meckel 1990: 166).

In differing ways, Peller, Baird and Potter shared the environmentalism common to medical thinking of the interwar years. Peller and Baird were situated in social medicine, which privileged social hermeneutics in the explanation of health and illness, particularly social class as a determinant of health, acting on the body through nutrition and housing. Potter was part of a generation of American physicians trained from the late 1920s who regarded hereditarianism as obsolete. Fetal and neonatal mortality was "only rarely due to diseases which are intrinsic to the fetus" wrote Potter and Adair (1943: 1065). As to Pfaundler, his first publications appeared at the beginning of the twentieth century. His writings demonstrated a pronatalism and concern for the conservation of population that marked the demographic concerns of this period, specifically the decline in national birthrates. But Pfaundler was forced into early retirement after the *Anschluss* (the German invasion of Austria during 1938), which the author of a memorial article on Pfaundler's work attributed to his lack of cooperation with the Germans after the invasion of Austria (Wiskott 1966: 104–105).

Peller's distance from eugenic thought stemmed from biographical and political sources. Peller was Jewish and had been subjected to anti-Semitism from his youth in Tarnopol (then part of the Austro-Hungarian Empire).[29] He was active in Palestine during the 1920s, establishing the Bureau of Public Health within the Zionist Executive there, serving as its first Director. A socialist, Peller was a member of the Austrian Committee of the *Eisenacher Verband*, a key group of intellectuals in the Austrian and German Socialist Parties, and was active in the mass campaign for the decriminalization of abortion in Germany (Peller 1931a; 1931d; 1967: viii; Weindling 1989: 461). After the banning of the Austrian Social Democratic Party in February 1934 and the destruction of parliamentary democracy, Peller, in common with other Jewish intellectuals who had ties to socialist organizing, escaped Austria in order to evade arrest. Peller had advocated welfare state policies, a common political position among those in social medicine.

Internationally, eugenic thought was not a fascist or a right wing monopoly, varying along party, ethnic, and class lines (Buchanan, Brock, Daniels and Winkler 2000), nor was it inherently hereditarian. Within Austria and Britain, for instance, social democracy and social medicine partially overlapped with eugenic thought. Those whom Daniel Kevles (1995: 173–176) has called "reform eugenicists" gave greater weight to social over hereditarian explanations of ill health, calling for improved wages, housing, diet, health care and education. During the late 1930s criticisms of German eugenics by members of the British Eugenics Society attracted to its ranks social democrats and those with an interest in social medicine, Richard Titmuss being one of these (Oakley 1991).[30] It is thus not startling that Baird, who was a personal friend and research collaborator of Titmuss (Oakley 1984: 310–312), participated simultaneously in social medicine and the British Eugenics Society, but it is curious that he should have remained with the Society into the 1970s, particularly to the extent of becoming an executive officer. How Baird came to serve as Fellow of the British Eugenics Society in 1944 and 1957, Director in 1971, and Vice-President from 1971 to 1974 remains a story that deserves to be told (see note 17 to this chapter). Baird's analysis of fetal and neonatal mortality in Aberdeen associated mortality with social stratification, explaining neither mortality nor social stratification in hereditarian terms. In his remarks given to a session on "Stillbirths and Neonatal Mortality" at the 1942 meeting of the Royal Society of Medicine, Baird pointedly rejected eugenic explanations of Scottish poverty and the high Scottish stillbirth rate ("Discussion on Stillbirth and Neonatal Mortality" (1942–1943): 68), disagreeing with Sir Francis Fremantle's (M.P.) remarks associating poverty and mortality with "natural weakness in breeding, character, or intelligence" ("Discussion on Stillbirth and Neonatal Mortality" (1942–1943: 64). With dry wit, Baird laconically pointed out that environment needed to be standardized for the question to be answered.

After its introduction by Peller into English, "perinatal mortality" came to be consistently used as a term of art in medical writings from the mid-1950s. The changes between the tenth (1950) and the eleventh (1956) editions of *Williams Obstetrics,* the standard textbook used in Canada and the USA for obstetrical training, can serve as a benchmark for this change (Hahn 1987). No reference to "perinatal mortality" appeared in the 1950 edition, which instead used the locution "total infant loss" for an estimate of mortality from stillbirths and neonatal deaths in the USA for 1946 (*Williams Obstetrics* 1950: 1). The 1956 edition introduced "perinatal mortality" to encompass fetal deaths after twenty weeks together with neonatal death in the first four weeks after birth, justifying the term "perinatal" on the basis of convenience: "Because the phrase "stillbirth and neonatal mortality" is a rather long and awkward term, the shorter description, perinatal mortality, is often used to encompass all deaths in these 2 categories" (*Williams Obstetrics* 1956: 12–13). Yet the 1956 edition of *Williams* still dismissed "perinatal mortality" as a concept lacking clinical value, awaiting better scientific understanding.

During the mid- to late 1950s "perinatal mortality" displaced the earlier notion of "stillbirths/fetal and neonatal mortality", with none of the English-language studies referring to Pfaundler or Peller. The *Oxford English Dictionary* cites a 1952 article in the *American Journal of Public Health* (Wallace *et al*. 1952) as the first use of "perinatal", although it had previously appeared in Peller's English-language articles (1948: 411, n. 1; 1944: 373, n. 10) in the *Bulletin of the History of Medicine*. The locution was taken up by American public health studies of the mid-1950s, appearing in Kohl's *Perinatal Mortality in New York City* (1955). Alfred Yankauer's "The Public Health Aspects of Perinatal Mortality" followed in 1957.[31] In England, Crosse and Mackintosh's (1954: 63) chapter, "Perinatal Mortality", in the collection *Recent Advances in Paediatrics*, began by noting: "[t]he term perinatal has come into use quite recently. It has been defined as the number of stillbirths and neonatal deaths per thousand total births (live and still)." They sought to explain the "factors" determining perinatal mortality, correlating it with parity (number of previous births), maternal age and social status. Crosse and Mackintosh were followed by Bound, Butler and Spector (1956a and b) who developed a classification of what they called "factors" leading to "perinatal mortality", subsequent to a preliminary gloss explaining the meaning of "perinatal mortality" to their audience. By the late 1950s perinatal mortality surveys had begun at national (England and Wales) and provincial (Ontario) levels, reporting their results in the early 1960s (Butler and Bonham 1963; Ontario Perinatal Mortality Study Committee 1961a and 1961b).

Like the prior division, "fetal and neonatal", "perinatal" was predicated on continuities between fetus and neonate. Neville Butler, co-author with Dennis Bonham of *Perinatal Mortality: The First Report of the 1958 British Perinatal Mortality Survey* (1963), on perinatal mortality summarized the commonalities between fetus and neonate in a 1966 report to the World Health Organisation :

> These two groups [fetus and neonate] are usefully combined for several reasons. Firstly, there is often a similarity between pathological lesions responsible for late foetal and early neonatal death. Secondly, in their social, biological and medical characteristics, early neonatal deaths are more aligned with late foetal deaths than with infant deaths between one week and one year. Thirdly, the premature foetus often survives labour but dies in the first few hours or days of life after withstanding the same degree of intrauterine oxygen-lack as results in death in labour of a mature or post-mature foetus. Fourthly, the effect of obstetric intervention, such as induction of labour or operative delivery, is best judged by combining late foetal and early neonatal deaths. Fifthly, there are still international variations in defining the signs of life.
>
> (Butler 1966: 1)

Butler justified combining fetal and neonatal mortality according to a number of quite separate criteria: similarities in causes of death, social characteristics, and bodily substance, together with a better standard for the evaluation of obstetrical services, and the lack of international agreement in what was considered a stillbirth. Baird, in his *Combined Textbook of Obstetrics and Gynaecology*, made the point more tersely: "it was a matter of chance whether an infant died shortly before or shortly after birth" (Baird 1962: 632). The formation of the "perinatal" as one continuous time interval eliminated the "chance" difference of birth.

Where the infant mortality rate had been defined in relation to deaths after live births, perinatal mortality and the earlier "fetal and neonatal mortality" combined live births with stillbirths. Those born dead and those dying early were joined in a common measure of potentially preventable mortality. Physicians came to contemplate a new antagonist in the struggle to conserve the health of population: "The number of lives lost in New York State during the five months extended from the twentieth week of gestation to the seventh day after birth is almost equal to the number lost during the next forty years of the human life span" (Yankauer 1957: 2499). The construction of the perinatal interval as a time span bridging before and after birth did not displace the distinction between stillbirths and newborns, however; no one proposed the recognition of "perinates": "There were 689 perinatal deaths, comprising 369 foetal deaths and 320 early neonatal deaths"

(Ontario Perinatal Mortality Study Committee 1967a: 23). The "perinatal" is a continuous temporal category from which the presence of birth has been simultaneously effaced and internalized as a distinction within a tripartite division of time into the antenatal, intranatal and postnatal, freeing the "perinatal" of the bodily presences found in the distinction between "fetal and neonatal".

By the late 1960s, the concept of perinatal mortality had been accepted internationally as a sensitive measure of maternal and child health. The eighth revision of the World Health Organisation's *Manual of the International Statistical Classification of Diseases, Injuries, and Causes of Death (ICD)*, a manual intended to standardize medical nomenclature globally, incorporated "perinatal mortality and morbidity" as a category (World Health Organisation 1967). It has been absent from the prior edition of 1957 (World Health Organisation 1957). At the 1965 Conference held prior to the eighth revision, delegates accepted the recommendation made in a 1963 report of the Sub-Committee on Classification of Diseases that the perinatal period be defined as lasting from the twenty-eighth week of gestation to the seventh day of life (World Health Organisation 1967: xxvii), and this was later officially adopted by the World Health Assembly.

Mortality rates are used to judge the health of population and the effectiveness of health care services; since health and health services are acknowledged to change, statisticians and medical researchers anticipate that new measures will be needed. The utility of infant and maternal mortality rates as measures of population health, had, it was reasoned during the 1950s and 1960s, been historically superseded in Europe and North America, and the new measure, perinatal mortality, better served as an index of maternal and child health:

> Progress in obstetric practice and improvements in the maternity services can still be judged in many developing countries by the maternal mortality rate, i.e., the number of maternal deaths per 1000 live births. With improving obstetric technique and general maternity care, the maternal mortality rate in most of the European Region is happily now so low that its use as a statistical criterion makes it singularly unrewarding ... The perinatal death rate, still very much higher than maternal mortality and more suitable for statistical analysis, brings into focus efficiency of both maternal care and the services for the newborn.
>
> (Butler 1966: 1)

> It is recognized that infant mortality is a good general index to the health status of any community. In recent years it has been appreciated that the perinatal mortality rate represents a more refined index of the care of mothers and their newborn infants than infant mortality per se.
>
> (Ontario Council of Health 1971: 11)

In the historical memory of medicine, mortality falls, followed in close pursuit by ever more refined scientific classifications of mortality rates: a thoroughly progressivist reading of history and a monumentalizing of medicine. But much can be learned about population power from the study of how health has been measured. The keeping of vital statistics forms a fine-toothed machine subject to perpetual redesign.

Problematization contain concepts and arguments that go nowhere; they proceed by conceptual excess. Perinatal mortality was one among many new concepts proposed to analyse the field called maternal and child mortality. Most of these concepts were eventually discarded. Peller (1936) for example had proposed "*Gerbürtstod*" [death at birth]: total maternal and neonatal mortality in the first week. Lilienfeld, Pasamanick and Rogers (1955) coined "reproductive casualty", for death and disease from conception into childhood. David Danforth's *Textbook of Obstetrics and Gynaecology* (1966: 197) proposed another concept, "fenatal mortality" to measure pregnancy loss from conception to seven days post-birth. Attending to these failed concepts offsets notions that our present could not have been otherwise than it is.

Was "perinatal mortality" equivalent to "fetal and neonatal mortality"? Certainly, they were counted in the same fashion. Yet differing techniques have been attached to the two concepts. "Fetal and neonatal" mortality linked to a specific programme of preventive techniques: better obstetrical skills during birth, testing and treatment for syphilis, prenatal care, and maternity benefits. The successor category, perinatal mortality, attached to risk-based prenatal care. The conjoining of the perinatal with risk was a form of governance entirely unknown to the prior period of the "fetal and neonatal".

Perinatal mortality and prenatal risk governance

During the late 1950s the new post-World War II analytic epidemiology, "risk factor" epidemiology, attached to the new concept of perinatal mortality. A cloud of risk factors drifted over the perinatal interval. Risk factor analysis was folded into the prior form of prenatal care that had existed since the 1920s, reconfiguring prenatal care as a standardized, risk-based regime serving the governmental objective of reducing perinatal mortality. In order to reduce regional and national perinatal mortality rates, each primary health care provider was to be supplied with a set list of risk factors to routinely assess women for risk at each prenatal visit. Risk techniques made possible not only the analysis of perinatal mortality and morbidity, but its care – "perinatal care" as it begins to be called at this time – setting down a vast and detailed apparatus of pregnancy risk governance, effective at both the population and individual levels.

William Arney (1982), in *Power and the Profession of Obstetrics,* documented and analysed an abrupt change in obstetrics during the late 1940s, the beginning of what he terms the "monitoring period", and it is to this era that risk-based prenatal care belongs. The prior period of obstetrics (1890–1945) was characterized, Arney argues, by a "technology of domineering control" such as anaesthesia and forceps, a spatial focus on the hospital, and attempts by obstetrics to institutionalize itself as a speciality controlling the meaning and organization of prenatal care and childbirth. In the post-1945 era, "Monitoring and surveillance of any aspect of birth and every aspect of the environment surrounding birth replaced classic, dramatic intervention in pregnancy" Arney 1982: 94). In the "monitoring period", the patient's body became a site for continuous data collection; outside the hospital populations were subject to surveillance; and an ecological model of health as environmental, social and emotional became authoritative. I situate the "monitoring period" within the governance of perinatal mortality.

The analysis of perinatal mortality in terms of risk factors coincided with the creation of a new object for epidemiology during the 1950s, chronic disease (Terris 1985; Susser and Susser 1996; Susser 1987), with the Framingham study of cardiovascular disease being the leading investigation in this new, analytic epidemiology (Dawber 1980). Declining rates of infectious diseases in the advanced capitalist world, combined with the mass availability of penicillin, promised the "conquest" of infectious diseases, an illusion of the rich that would not be displaced until the end of the twentieth century. Post-World War II epidemiology witnessed a general lightening of explanation in terms of single pathological or social causes towards analysis of the relations between "exposures" and "outcomes" in multifactorial/multicausal models of disease (Pearce 1996: 680; Susser 1987: 31–35). These exposure–outcome relations were named risk factors; the intelligibility of risk factors lay in their being a measure of association rather than a cause: the probability of an outcome from an exposure. Risk factor epidemiology formed a particulate plane of tiny concepts, a surface that displaced the depth explanations of social medicine, which treated illness as caused by social structures and processes. The perinatal mortality surveys of the late 1950s and 1960s were designed in terms of the new analytical epidemiology that construed signs, symptoms, behaviours and even diagnoses as risk factors increasing the likelihood of particular outcomes (Armstrong 1995: 400). In this process clinical care during pregnancy became oriented to the risk of future pathology, alert to anticipations of what might happen, attentive to pre-symptomatic indications (Armstrong 1995).

Despite the continuing hold of risk-based prenatal care over our present, it is based on a form of epidemiology that has been widely contested in the health sciences. From the 1980s, epidemiologists came to criticize "risk factor epidemiology" on a number of grounds. It was said that risk factors were heuristic devices having little theoretical basis, taken either singly or in relation to each other, leaving epidemiology with no conceptual basis. So too, lists of risk factors supposed each factor to be of equal significance in producing the outcome of interest, an assumption blind to the question of which factors might be determinate (Krieger 1994). Lastly, epidemiologists noted the pervasive methodological individualism of risk factor epidemiology, which assumed a population to be the sum of the individuals comprising it, and risk as an inherent property of the individual (Jackson 2005: 82–84). During the 1970s and 1980s the health promotion model was articulated to risk factor epidemiology, which resulted in a model of preventive health dedicated to the identification of the "high risk" individual and the inculcation of individual responsibility for risk reduction, together with a reciprocal disinterest in the determinants and comparative analysis of disease in populations (Jackson 2005: 76–86). Despite the many internal medical critiques of risk factor epidemiology, the risk-based system of prenatal care that was first conceptualized in the 1950s continued unabated, though questioned, into the late twentieth and early twenty-first centuries.

During the mid-1950s a number of attempts were made to form a classification of perinatal deaths by cause (Baird, Thomson and Duncan 1953; Baird, Walker and Thomson 1954; Bound, Butler and Spector 1956a and b; Kohl 1955; Yankauer 1957). The classifications were internally heterogeneous, based on the disparate forms of medical reasoning that had been active in problematizing "fetal and neonatal mortality": vital statistics, pathology, and clinical reasoning. The task was made difficult by the profusion of types and numbers of factors involved. To begin with, the temporal division of factors for perinatal mortality was threefold: gestational, intrapartum and postpartum. Maternal factors needed distinguishing from fetal factors. Perinatal deaths were also associated with both social and pathological factors. The identification of risk factors for the reduction of perinatal mortality was driven by an implacable inductive energy to be utterly exhaustive in the inclusion of risk factors.

Butler and Bonham's internationally cited study, *Perinatal Mortality* (1963), was based on a British national survey of 1958. A private foundation, the National Birthday Trust, funded the research and report. The purpose of the research was to identify and quantify risk of perinatal death according to a wide range of social, pathological, medical, biological, and clinical factors (for discussion see Tew 1990: 238–245). Perinatal deaths were analysed by geographical region, maternal age, parity (number of previous pregnancies), social class (maternal), past obstetric history, place of booking and prenatal care, toxemia and bleeding in pregnancy, gestation and birth weight, and labour and "delivery", postmortem identification of cause of death. These were referred to as "factors" for the most part in the Butler and Bonham study, although occasionally "risk" and "perinatal mortality risk" also appeared. *Perinatal Mortality* correlated these factors for the periods before, during and after birth, in certain cases measuring the association of more than one factor. In its pages the combinatory logic of analytic epidemiology constituted a flat associative plane making obstetrical history, residence and clinical signs into instances of a higher order category: factors.

Perinatal Mortality formulated a framework for the application of risk to prenatal care. The 1963 Butler and Bonham report recommended a series of changes to reduce perinatal mortality, calling for "better selection of women for hospital prenatal care and confinement, earlier recognition and treatment of toxemia in pregnancy, greater recognition of the importance of length of gestation and above all, unification and extension of the Maternity Services" (Butler and

Bonham 1963: xi). Butler and Bonham conceptualized a reduction in the perinatal mortality rate to require both a clinical and an administrative solution.[32] Primary health care providers needed to screen women for place of birth (home or hospital) in terms of standardized risk factors, and to diagnose and treat other factors such as toxemia (high blood pressure) more effectively. The reduction of perinatal mortality was also thought to require a net of co-ordinated maternity services that would extend in principle to all British women.

By the mid-1960s Butler had projected the "high-risk mother" as the chief source of perinatal mortality and thus the primary target of health care with respect to perinatal mortality reduction:

> It is stressed that the major proportion of perinatal mortality comes from a vulnerable minority group of "high-risk" mothers who can, in many instances, be singled out on the basis of simple criteria so that the best available delivery and newborn facilities can be reserved for them.
>
> (Butler 1966: 20)

The problematization of perinatal mortality had found a simple solution in the high risk mother, a medico-administrative subject constituted from the clustering of risk factors. A programme of intervention around maternal risk levels proliferated over the next several decades, even after the singular category of the "high-risk" mother went into disuse during the mid- to late 1970s (Arney 1982: 143).

By the mid- to late 1960s the host of "factors" that had been attached to the knowledge of perinatal mortality during the late 1950s and early 1960s – "social factors", "maternal and infant factors", "gestational factors", "fetal factors", "prenatal factors", "intrapartum factors", "postpartum factors" came to be named *risk* factors, marking the consolidation of a cognitive transformation: the folding of risk into prenatal care. Groups of factors that had been located in previously conflicting forms of medical knowledge were placed contiguously; social factors and pathological factors lay down together in the peaceable kingdom of risk.

The period of the initial perinatal mortality surveys during the 1950s and early 1960s coincided with medical concern about variations in the quality of care patients were receiving from physicians and a corresponding attempt to standardize physicians' clinical judgement (Berg 1995). A risk-based prenatal care regime was produced by linking the drive to standardize clinical care with risk factor epidemiology. By the late 1960s a new technique for the reduction of perinatal mortality had been devised: risk screening of pregnancy through standardized printed forms. The sets of factors elaborated in the perinatal mortality surveys were recast as a device for risk reduction in routine clinical care. Factor analysis was made into a proactive screening technique focused on identifying the "high-risk mother" for the purpose of a more intensive regime of care. This comprises an instance of what I have called clinical risk: the application of risk, an epidemiological concept pertaining to population, to the bodies of individuals in clinical care. I return to the discussion of clinical risk at the conclusion of this chapter.

During the last three decades of the twentieth century standardized prenatal risk screening came into use in a number of Euroamerican health regimes, including Australia, Britain, Canada,[33] France, Germany, and the USA. The Ontario Antenatal Record (see appendix a, b and c to Chapter 3: 112–114), a risk assessment tool used in the Canadian province of Ontario since 1980, comprised one of these screening tools.[34] Examining its history displays how risk governance came to organize prenatal care.

In Ontario, prenatal risk assessment is situated in a history of attempts to reduce mortality that began in the late 1950s. The Deputy Minister of Health for Ontario, Dr W.G. Brown, called a conference in November 1957, "to discuss some of the problems concerning stillbirths and early neonatal deaths" (Ontario Perinatal Mortality Study Committee 1967a: 1). The 1957 Conference resulted in the formation of the Perinatal Mortality Study Committee which went on to perform a survey of perinatal mortality in Ontario teaching hospitals. The reports of the 1959 survey (Ontario Perinatal Mortality Study Committee 1961a and b, 1967a and b) focused on the "factors" with which perinatal mortality was associated. The sets of factors analysed were similar to that found in Butler and Bonham, although with some variations, such as the addition of maternal height. The Ontario reports were characterized by an absence of interest in social class other than in the weakened form of "employment in pregnancy", a pointed contrast to the analytic significance of social class in Butler and Bonham's *Perinatal Mortality*. No tests of significance were performed on the data, with the result being long and diffuse sets of suspect factors whose association with the likelihood of an "adverse outcome" was uncertain. The first Report of the Perinatal Study Committee (1961a: 18) concluded by calling for "more information" to evaluate the following: "prenatal care, previous obstetrical history of abortions or premature delivery, marital status, mothers cared for in public wards, blood grouping (ABO and Rh), blood pressure over 140–80, hydramnios, hyperemesis, diabetes" together with another eleven factors ranging from care during labour to maternal injury during pregnancy. The intent of the survey research was to compose a comprehensive list of all prenatal, intranatal, and neonatal risk factors, a list that could be used to provide the basis for perinatal risk reduction. However, the 1961 and 1967 *Reports* contained no specific recommendations about how exactly the reduction of the perinatal mortality rate was to be accomplished. The elision of the distinction between risk and diagnosis in the 1961 and 1967 *Reports* was to characterize the future system of risk-based care.

The next in the series of Ontario reports, *Perinatal Problems* (Ontario Council of Health 1971), targeted economic, regional and between-hospital variation in perinatal mortality by recommending the establishment of high-intensive perinatal units across the province. *Perinatal Problems* proposed that women be referred to these high-intensive units for care on the basis of an assessment that used a standardized risk screening form. The Report recommended that "all prenatal patients" (Ontario Council of Health 1971: 3) be risk screened in order to identify a subgroup of pregnant women, "the high-risk mother", who, it was argued, were the chief source of preventable perinatal mortality and morbidity (Ontario Council of Health 1971: 12).

Perinatal Problems was the first report published by the Ontario Department of Health recommending the use of standardized risk assessment in prenatal care on a population basis. The Report contained a prototype risk assessment form (Ontario Council of Health 1971, Appendix A, "A Scoring Index for High Risk Pregnancy Assessment": 33–36).[35] The perinatal risk factors listed in the two prior Reports of the Perinatality Mortality Study (Ontario Perinatal Mortality Study Committee 1961a and b; 1967a and b) became the basis for the factors appearing in the prototype for risk scoring.

The 1979 Report of the Advisory Committee on Reproductive Medical Care to the Ontario Minister of Health repeated the recommendation for uniform risk assessment made in *Perinatal Problems* (Ontario Council of Health 1971), asking that the "Antenatal Record approved by this Committee [the Advisory Committee on Reproductive Medical Care] be mandated for use in Ontario" (Advisory Committee on Reproductive Medical Care 1979: x; see Appendix D, n.p., for model Antenatal Record). A revised version of the Antenatal Record was approved by the Ontario Medical Association in 1979, coming into provincial use during 1980.[36]

The first generation of standardized prenatal risk assessment forms were developed during the 1960s and 1970s. These early risk scoring devices gave each factor a numerical value; the numbers were then added together to form a "risk score" for a woman in care (Alexander and Keirse 1989: 345–346).[37] During the 1970s a variety of prenatal risk screening forms were used in Ontario, though no form was accepted as standard. The best known of these risk scoring forms was developed by James Goodwin (Goodwin, Dunne and Thomas 1969). Goodwin, Dunne and Thomas (1969: 57) argued the need for prenatal risk scoring on the basis of Canada's (dismal) twelfth place rank among Western nations with respect to its perinatal mortality rate; they attributed this poor showing to "our failure to recognize the fetus at risk". The article identified a series of "fetal-risk factors" using "data from perinatal mortality surveys" (citing Butler and Bonham 1963 and Ontario Perinatal Mortality Study Committee 1967) combined with their own clinical experience (Goodwin, Dunne and Thomas 1969: 57).

For Goodwin, Dunne and Thomas (1969) the main objective of perinatal risk scoring consisted of reducing mortality to the "high-risk fetus": a displacement of the "high-risk mother" as the principle subject of perinatal risk measures. But, despite appearances, the identification of the "high-risk mother" had not aimed at improving the health outcomes of mothers, but in reducing risk to the unborn. Whereas prenatal care prior to 1960 mainly attempted to benefit maternal health, the introduction of risk into prenatal care was connected to reducing perinatal mortality, and was thus about conserving fetal and neonatal health.

Ontario provincial reports studying perinatal mortality during the 1970s worked in the context of a general move to regionalize health care in the province.[38] Based on size and technical capacity, health units were categorized in terms of three levels of intensity – a streaming of health care delivery from more to less specialized. Ontario was spatially partitioned into regions, with the most complicated and costly types of care made available in the centre of the region. Regionalization addressed what had been identified as provincial inequalities in access to care. *A Regionalized System for Reproductive Medical Care in Ontario* (1979) mapped the tertiary care model onto prenatal care, dividing pregnancies into three risk levels corresponding to the three levels of regionalized care.[39]

The 1979 *Report* clustered risk factors into three groups ranked from low to high levels of risk. The harmonizing of prenatal risk levels with levels of intensity of care had first been proposed during the mid- to late 1960s; risk governance and regionalization had in effect grown up together. In the words of Barbara Perkins, "relationships between risk and structure have been reciprocal" (Perkins 1994: 26). The reduction of the perinatal mortality rate, implicated as it is in a tertiary care model, is economically costly, much more so than the reduction of the infant mortality rate.

The first appearance of the Ontario Antenatal Record was announced in an article published in the *Ontario Medical Review*, a publication of the Ontario Medical Association mailed to all members of the College of Physicians and Surgeons of Ontario. The first author, James Goodwin, was well known for his work (described above) in designing a "fetal risk" scoring technique used in Canada and the United States (Goodwin, Dunne and Thomas 1969). The second author, Graham Chance, was a member of the Ontario Medical Association Special Committee on Perinatal Care, which had taken on the work of co-ordinating the design of a standardized prenatal risk assessment device. The first paragraph of Goodwin and Chance's article squarely located the Ontario Antenatal Record within the history of perinatal care and fetal risk assessment:

> The development of modern perinatal care over the past 20 years has unquestionably reduced fetal and neonatal mortality, and with it, significant illness and brain damage in the newborn. Clearly, sound perinatal care begins with the identification of those fetal and maternal patients who are at risk. All pregnancies may be subject to unpredictable hazard and should be considered potentially at risk until proven otherwise. Since 1969, fetal risk scores of varying complexity have been proposed in an attempt to solve the thorny problems of identification ... This paper presents a new proposal for a combined antenatal record and fetal risk assessment at each antenatal visit. It is hoped that this fetal risk assignment system will serve as a basis for the practical management of the high-risk pregnancy.
>
> (Goodwin and Chance 1979: 563)

The intent of the Ontario Antenatal Record marked the institution of a "practical fetal risk assessment system" (ibid.). The resulting form had the dual function of serving as a record of pregnancy and as a fetal risk assessment system. The authors noted that the "assignment of risk-grading should necessarily involve an assessment of maternal as well as fetal hazards" (Goodwin and Chance 1979: 564). Although the subject of risk assessment was primarily the fetus, it also interpellated the pregnant woman, but the relation between fetal assessment and maternal assessment was unclear, as was the relation of "maternal factors" to "fetal factors" (Hall 1994: 1241–1242). Risk-based prenatal care has a pervasive ambiguity as to the subject(s) of care.

First used province-wide in 1980, the Ontario Antenatal Record has come to be widely accepted, recommended for physician use by the Ontario Medical Association. Neither the Ontario Medical Association nor the provincial Ministry of Health require its completion, but hospital boards do.[40] Since 1998 all Ontario hospitals doing obstetrical care have required its submission for women giving birth on their premises. The Ontario Antenatal Record is not processed for official statistical purposes provincially or federally, although the completed forms are used in perinatal death and morbidity reviews done at a local hospital level (Hanvey *et al.* 1991). The impact of the Ontario Antenatal Record has thus mainly been at the clinical level. Interestingly, there is no comparable form for any other health condition issued by the Ministry of Health or the Ontario Medical Association that requires such detailed recording of clinical practice. Pregnancy and childbirth are the object of a unique textual process.

Particular risk factors in prenatal risk assessment are over time deleted, others added; there is contestation over the appropriateness of individual factors; international and regional differences occur. Yet prenatal risk assessment as a technique persists across these variations because it operates according to a schema that acts as a higher-order guide to action – a set of discursive distinctions at the level of governance. The schema of prenatal risk assessment consists of a standardized list of risk factors applied in clinical practice on a population basis in order to isolate those at higher risk and to act on that risk preventively.

I should note that the account of risk-based prenatal care offered here has not focused on what are called "modifiable risk factors". Much has been written about the attempt to "responsibilize" women during pregnancy to control exposure to risk factors such as smoking, alcohol use and the consumption of prohibited drugs. More broadly, it has become something of a convention to associate advanced liberal forms of governance with inciting the "active citizen" to reduce his/her exposure to risk. Such risks are known within epidemiology as "modifiable risk factors": a determinant (an attribute or exposure that increases the probability of an outcome/disease) that can be modified through intervention.[41] Risk factors, however, also include determinants that cannot be modified by either a clinician or the person in care.[42]

The genealogy of prenatal risk found here has shown a marked disinterest in modifiable risk factors. Reducing risk to modifiable risk factors oversimplifies the organization of risk within health knowledges. Modifiable risk factors comprise only one aspect of prenatal risk assessment, and of these only a small number are modifiable by the pregnant woman. Risk factors such as parity (number of previous births), height and past reproductive history cannot be modified. None the less, modifiable risk factors have particular power/knowledge significance in linking women's actions during pregnancy to the governmental objective of lowering perinatal mortality and morbidity rates. Those who do not govern themselves so as to limit their risk exposures may be perceived as undermining population health and injuring future generations. These contraventions have elicited a long line of sanctions, notably at law (a topic investigated in Chapters 4 and 5).

This microgenealogy of the Ontario Antenatal Record has situated the history of risk-based prenatal care as a response to the question of how to lower perinatal mortality within the broader context of population politics. Contextualizing prenatal risk screening within the history of perinatal mortality reduction produces an account of how and why pregnancy and risk governance came to be articulated. Prenatal risk screening comprised a clinical technique for a governmental end: the conservation of population through the reduction of perinatal mortality. In the risk-based prenatal regime, the care of the fetus so as to prevent perinatal mortality became a routine clinical accomplishment.

The instability of clinical risk: a note on risk governance

The genealogy of prenatal risk assessment presented here assists in conceptualizing the specificity of risk reasoning at the site of the clinical encounter. Prenatal risk assessment is an instance of what Sandra Gifford (1986) has termed *clinical risk*: the application of an epidemiological knowledge of risk in populations to the clinical care of individual patients. Clinical risk conjoins two conceptually distinct forms of health judgement: the judgement of risk (from epidemiological reasoning) and the judgement of the normal and the pathological (diagnostics). I argue that clinical risk aggregates heterogeneous and incommensurable forms of health reasoning. It is thus analytically incoherent and intrinsically unstable. Although this incoherence has been recognized and repeatedly contested within medicine and the allied health professions, uses of risk reasoning in clinical practice persist, the internal critique constantly forgotten. The system of this disorder and its functions require investigation.

Epidemiology studies populations, while clinical medicine involves the diagnosis and treatment of persons. Whereas in epidemiology risk from an exposure is a quantitative concept, calculable over a study population (see Figure 1.1), in clinical care risk from an exposure to an individual patient is not calculable. Technically speaking, risk in clinical care is uncertain; the outcome from an exposure may be positive or negative, and this can only be known in the future, not at the time of care. That is, it is undecidable whether a patient with a risk factor should be assigned to cell a or c of Figure 1.1. Thus clinical risk does not strictly speaking conform to the same grid of analytic intelligibility as epidemiological risk. A field of tensions runs between risk judgement in population-based epidemiological knowledge and the pronouncement of the individual as healthy or sick in clinical care. Risk suffers from an intrinsic instability in clinical care since the risk probabilities that apply at a population level do not predict individual patient outcomes.

Clinical risk is one of numerous and interesting forms of governing through uncertainty found in our present, particularly the precautionary principle: a logic of vigilance and pre-emptive action in order to eliminate events regarded as catastrophic (Ewald 2002). But clinical risk interestingly codes incalculable uncertainty as calculable risk. Indeed there are other senses of "risk" in contemporary medicine that also have the property of being technically uncertain. These often occur in the field of global health, such as in the title of the 2004 Conference, "Controlling the Risk: Science to Combat Global Infectious Diseases", where risk meant something like danger/threat.[43] Risk borrows authority from epidemiology; it bears the scientific accent of epidemiology.

When epidemiological risk is transposed to clinical care, the resulting risk technique is marked by the imbrication of clinical concepts and diagnostics. The lists of risk factors for prenatal screening include clinical signs, symptoms, behaviours, test results and even diagnoses. The edition of the Ontario Antenatal Record (see reproduction in the appendix a, b and c to Chapter 3: 112–114) that was used during the 1990s and first years of the twenty-first century illustrates this concatenation of disparate categories. In the "Guide to Pregnancy Risk Assessment" diagnostic categories appear as risk factors, e.g. "diabetes", "renal disease without hypertension" and "other significant medical illness". No clear demarcation between risk and diagnostics is discernable in the Ontario Antenatal Record. In the elision between risk and diagnostics, risk factors proliferate and the state of health becomes elusive, a condition of "no predictable risk", that is, uncertainty. Risk encroaches on the meaning of health, almost displacing it (see the appendix a headed "Risk Level A. Healthy Pregnancy, no predictable risk").

Clinical reason conceives diseases as immanent in the body of the person, taking variable courses through the individual body, courses that the clinical gaze must discern through interpreting signs, symptoms and diagnostic tests. A series of studies following from Sandra Gifford's important 1986 article have suggested that, when epidemiological knowledge is inserted into the clinical setting, risk factors are read as equivalent to "objective clinical signs of disease" (Gifford 1986: 222). Practitioners in clinical care equate risk factors with clinical signs and symptoms located in the patient's body, as Adelswärd and Sachs (1998: 200) note forcefully: health care providers "seem to locate risk within, and to make risk a property of the individual instead of constructing risk as one aspect of a complex, hypothetical future event involving numerous interrelated aspects, whereby the event cannot be located precisely – neither in time, space, nor body". Patients receive treatments for risk factors on the grounds that eliminating the factors will improve their prognosis. In clinical practice both physicians and patients interpret risk as a state *between* health and illness (Gifford 1986: 215; Robertson 2001 and 2000: 65–67; Adelswärd and Sachs 1998: 205). Chapter 3 of this book examines how midwives orient to the task of obligatory pregnancy risk screening, a task at odds with their commitment to the interpretation of pregnancy as physiologic, as a state of health.

The effects of binding risk to diagnostic categories and test results have been criticized *within* medicine and the allied health professions, principally from research-based and evidence-based health sciences. The inclusion of diagnostic elements within prenatal risk assessment has been contested on the grounds that it simply records what would already be in a patient's chart, inflates the reliability of the risk tool, and initiates a course of care that would have occurred without the assessment tool (Alexander and Keirse 1989: 347–348). No strict boundary exists between diagnostic and risk categories in the schema of risk assessment: risk has the property of binding to diagnostic judgement, transforming it into risk factors. In this process, risk assessment borrows reliability from clinical medicine.

It is likewise commonly observed within the health sciences that the presence of a risk factor does not necessarily result in an individual's developing the condition which s/he is at greater probability of acquiring. The removal of a risk factor is recognized as potentially having no impact on the health of individuals (Sackett *et al.* 2000: 69). But, since risk factors tend to be equated with clinical signs, this results in patients being treated for risk factors (Gifford 1986: 222). Thus risk screening has been criticized as giving rise to unnecessary interventions and iatrogenic effects.[44] The work of David Sackett and his colleagues in evidence-based medicine has been strongly critical of risk screening for its iatrogenic effects:

> [A]ll screening and case-finding, at least in the short-run, hurt people. Early diagnosis is just that: People are "labeled" as having, or as being at high risk for developing, some pretty awful diseases (cancer of the breast, stroke, heart attack and the like.) And this labeling takes place month, years, or even decades before the awful diseases will become manifest as symptomatic illness (often in only a small portion of those who screen positive). Labeling hurts. For example, a cohort of working men studied before and after they were labeled hypertensive displayed increased absenteeism, decreased psychological well-being, and progressive loss of income in comparison with their normotensive workmates ... What's even worse is that those with false-positive screening tests will experience only harm (regardless of the efficacy of early treatment). But even individuals with true-positive tests who receive efficacious treatment have had "healthy time" taken away from them; early diagnosis may not make folks live longer, but it surely makes all of them "sick" longer!
>
> (Sackett *et al.* 2000: 90)

The literature of evidence-based medicine has been cautious and technical about the transposition of research based on population studies into clinical care.

The line of argument in which I am engaging has not attempted to discredit clinical risk as false, but instead to show the existence of critiques that point to competing modes of reasoning about risk within medicine and the health sciences. The critiques were of help to me in coming to an understanding of clinical risk as a conceptually incoherent aggregate of reasoning rather than, as I had first presumed, coherent and singular.

Historians of the present have analysed risk as the antithesis of diagnosis, its competitor and successor. Robert Castel (1991 and 1981), whose work has been foundational to the study of risk in advanced liberalism, has analysed the risk techniques of late twentieth-century psychology and criminology in terms of an historical shift from dangerousness to risk. Whereas nineteenth- and early twentieth-century psychology and criminology had targeted the identification and treatment of *existing* "pathology" in individuals, in the late twentieth century psychology and criminology became *predictive*, seeking to arrest future harm before it happened. While the earlier techniques for identifying the "dangerous" individual had aimed at treating his/her purported pathology, late twentieth century risk techniques scan client databases with risk profiles, following up with home visits to *potential* offenders.

In tandem with Castel, David Armstrong has argued that risk represents a break from prior forms of medical reasoning that were organized around the distinction between the normal and the pathological (Armstrong 1995: 395–397); risk systematizes the possibility of bodies neither normal nor pathological, bodies "at risk". The two modes of reasoning differ temporally and spatially. Diagnostic judgement pronounces the patient sick or well in the moment of its enunciation. Risk judgement orients to the future, giving the clinical body increased temporal depth. In clinical reasoning pathology is conceived as a site of signs and symptoms explicable in terms of processes occurring in the depths of the body, contrasting with risk judgement as a mode of reasoning about prevention through control of risk factors. These risk factors may belong to the space of the clinical body, but may also be external to it, located in "environment" or "lifestyle" (Armstrong 1995: 400–401). Risk operates on surface associations, not hermeneutically: it has no depths, no underlying causes. Clinical risk is part of a wider tension between surface and depth explanations found in medicine today such as the displacement of the case report by the news report as the unit of global infectious disease surveillance and the rise of syndromic surveillance measures that link sales data with outbreak potentially before a single clinical/laboratory diagnosis has occurred.

The analytic distinction between risk and diagnostics is clearly necessary, but the use of antithesis in drawing the distinction has occluded the ways in which they co-occur and interact. Prenatal risk assessment is a composite screening schema, formed from the assembling of comprehensive, standardized classifications of fetal and neonatal death by cause/factor during the period from the mid-1950s to mid-1960s. This work integrated prior forms of medical reasoning about fetal and neonatal death into a framework inclusive of social, pathological, clinical and physiological factors/causes. The causes/factors were folded into risk assessment at a clinical level during the late 1960s and 1970s. The drive to lower perinatal mortality through the reduction of antenatal risk aggregated disparate forms of medical reasoning. Analytically prenatal risk assessment combined statistical, epidemiological reasoning oriented to the study of populations (e.g. the work of Peller and Baird on the association of perinatal mortality with socioeconomic status), with diagnostic and aetiological reasoning (e.g. the work of the pathological anatomists Holland and Potter in establishing cause of death). Diagnostics and risk, distinct forms of reasoning, were glued together in prenatal risk assessment, their relation blurred.

In the preceding discussion I have called attention to the significance of analysing the amalgamation of the incommensurable. Clinical risk is a space of tension, characterized by intrinsic instability, continually dissolving and being reassembled. It seams together two forms of knowledge: epidemiological and clinical reasoning. Analytically clinical risk combines two differing forms of health judgement: the judgement of risk and the judgement of the normal and the pathological. This constitutes an aggregation of the heterogeneous and the incommensurable. This system of disorder has persisted since the late 1950s, its effects continuing into our present, subject to a continual forgetting of its critique.

The perinatal threshold

The perinatal threshold of the living subject was instituted in routine clinical practice through standardized prenatal risk screening, that is, the projection of risk onto pregnant women's bodies. Prenatal risk screening, as I have shown, aimed to reduce perinatal mortality. Initially it is puzzling that perinatal and prenatal would be associated in this fashion, as "prenatal" care formally pertains to a woman during pregnancy, as opposed to "perinatal care", which is care of the fetus and newborn (see the second epigraph to this chapter). The linking of the prenatal and the perinatal occurred through the attachment of perinatal risk factors to the system of risk-based prenatal care that was being formed from the mid-1950s. The previous form of prenatal care that existed from 1920 to 1950 had been concerned with maternal health and not organized through risk governance. The perinatal, and thus the care of the fetus and newborn, was folded into the prenatal at the invention of the risk-based prenatal care regime during the 1950s and 1960s. In announcing the distribution of the first Ontario Antenatal Record, the physicians who had designed it made clear that the Antenatal Record was intended to promote "the care of the fetus":

> With the introduction of the concept of perinatal risk assignment embodied in this new antenatal recording system, we hope that family physicians and obstetricians will feel an increased sense of participation in the care of the fetus and the assessment of fetal health.
>
> (Goodwin and Chance 1979: 564)

The security of the fetus late in pregnancy is thus a systemic effect of perinatal risk assessment found within prenatal care. Earlier attempts to lower the infant mortality rate in Ontario during the twentieth century had been enacted through mass education campaigns for women during pregnancy (Arnup 1994; Comacchio 1993), prenatal care (Mitchinson 2002), and a policy of what was richly called "universal hospital confinement" during childbirth (Oppenheimer 1990). Perinatal risk assessment comprised a policy of higher-intensity clinical care for women during pregnancy. While prenatal care in the first half of the twentieth century tried to secure the mother from the "complications" of pregnancy and childbirth, from the 1960s it focused on the care of the fetus through the assessment of perinatal risk factors. The new goal was to reduce perinatal mortality; the target was "high-risk" pregnant women; the means used were risk screening and specialized care in perinatal units.

The present analysis situates the emergence of the fetus in clinical care within a biopolitics of population at the perinatal threshold. It is reciprocally a genealogy of the threshold of the living subject in biopolitics, which, during the twentieth century, was extended prior to birth for the purposes of reducing perinatal mortality and morbidity. The "care of the fetus" in perinatal risk assessment formed one of the practices that pushed the conservation of population before birth.

The concepts of perinatal mortality and morbidity are medico-administrative categories forming constituents of translocal hospital, regional, national and global statistical systems. Perinatal mortality is thus located as a category within the vital statistics of population. Population, as Bruce Curtis (2001: 4) has forcefully reminded us, is an abstraction rather than an empirically existing object, an abstraction contingent on the existence of an authoritative collectivity that renders subjects equivalent with respect to criteria such as birth, death, and marriage. These abstract criteria occur in an administratively defined time (such as the year for annual statistics) and space (the territory in which the relations of governance obtain) (Curtis 2001: 26). The subjects rendered equivalent in official perinatal statistics are stillbirths and newborns, which exist in an annual administrative time correlated with a medical time (twenty-eight weeks' gestation to seven days after birth). These temporal divisions apply within an administrative space: the territory of the nation, region and hospital. In the first twenty-seven weeks of gestation the fetus belongs to no sovereign population, entering national mortality statistics only if dying after the twenty-eighth week. A noteworthy and pleasing administrative asymmetry currently exists between vital and census statistics with respect to perinatal deaths; stillbirths from twenty-eight weeks comprise part of vital statistics, but the unborn during the last trimester is not included in national censuses of population.

In 1959 the Committee on Maternal and Child Care of the American Medical Association published *A Guide for the Study of Perinatal Mortality and Morbidity*. With the flat and potentially malevolent naivety characteristic of this historical period, the Committee on Child Welfare clarified the purpose of the *Guide*: "The objective of perinatal mortality and morbidity studies is to improve the production of normal human beings" (American Medical Association 1959: 5). To this end, the Committee proposed a standardized definition of perinatal mortality for statistical purposes to be used at the hospital, state, and county levels. The definition would, the Committee suggested, act in tandem with a proposed infrastructure of hospital, city, and state "perinatal mortality and morbidity study committees", thus co-ordinating an administratively defined time and space within relations of governance. The optimizing of health at the perinatal threshold is pervasively mediated through sovereign and governmental textual practices that operate to concert and standardize the reporting of local particularities.

Seeking to stabilize statistical interpretive practices for counting perinatal mortality and morbidity, the Committee on Maternal and Child Care provided a "Perinatal Period Chart" (see Figure 2.3). This chart gives a representation of the perinatal threshold as the projection of a standardized medico-administrative time for the statistical purpose of counting a population, assisting a larger and grander project: "the production of normal human beings". Although the Committee mentioned birth in its definition of perinatal mortality as "those deaths of fetuses and newborn infants occurring before, during, and soon after birth" (American Medical Association 1959: 7), birth does not appear as a category on the chart. Yet this representation of the perinatal threshold is bilaterally symmetrical, with birth as its mid-point; the vertical axis designates the birth process, abstractly symbolized by ")↑(", which suggests what physicians and midwives call the "birth canal".

Birth is both present and absent at the perinatal threshold: a moment of transition from fetus to neonate, a moment of pure verticality dividing the horizontal axis of physiological development that links unborn and born. To the left of the vertical axis appear fetal weights, increasing from 500 grams to 2,500 grams (and over) prior to ")↑(" (birth). The time of pregnancy is absent from the perinatal threshold, displaced by fetal weight. For the statistical purposes of the American Medical Association fetal weight has greater reliability than gestational age in counts of stillbirths. To the right of the vertical axis appear integers representing the first seven days of infant life (the neonatal period), extending at the far right to twenty-eight days. Birth is thus constituted as a caesura articulating fetal weight to neonatal time.[45] Two perinatal periods are found in "The Perinatal Period Chart"; Perinatal Period I starts at 1,000 grams and ends at seven days after birth, and Perinatal Period II at 500 grams and twenty-eight days after birth. The temporal interval of the perinatal can be expanded or contracted depending on medicine's choice of its antagonist in reducing rates of death and illness; the duration of the perinatal interval is not fixed.

Although "The Perinatal Period Chart" constituted an effort on the part of the American Medical Association to standardize statistics on perinatal mortality, the invention of the perinatal threshold involved far more than simply counting mortality after the fact. The perinatal threshold in part comprised an attempt to prevent stillbirths from occurring through a variety of measures such as routine risk-based prenatal care. The perinatal threshold was a medico-administrative

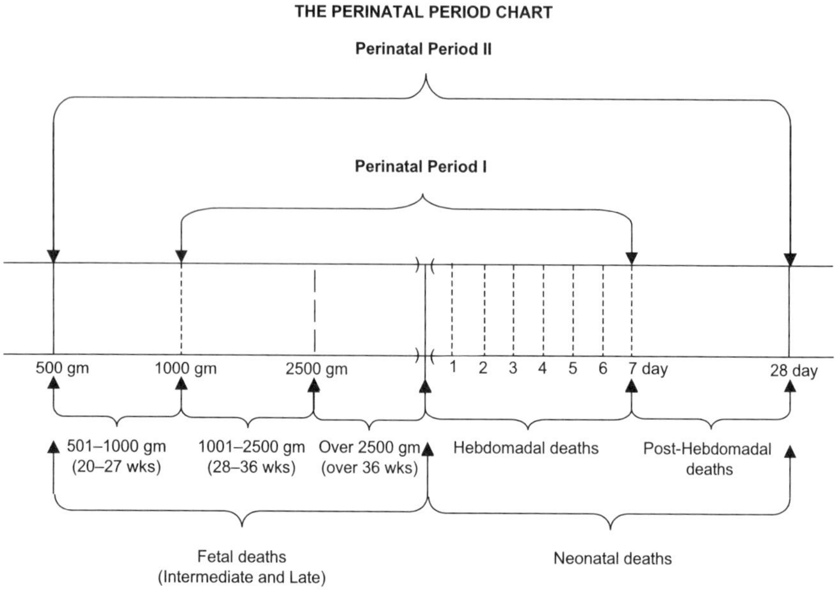

This is a reproduction of "The Perinatal Period Chart", a work of the Committee on Maternal and Child Care, American Medical Association. This chart comprised part of an attempt to standardize the definition of perinatal mortality and to enlist physicians in reporting it. The "Perinatal Period Chart" represents the perinatal threshold as the projection of a standardized medico-administrative temporal interval used for the statistical purpose of counting a population. Note the unnamed representation of birth as ")↑(", the vertical axis in the chart as a moment of developmental transition linking fetal weight to neonatal time. The object of the perinatal threshold is the combined fetus-neonate given in a mode of medico-administrative vision that reads pregnancy as a sequence of fetal weights and labour as a vertical axis.

Figure 2.3 The perinatal threshold: a representation

Source: American Medical Association (1959) Committee on Maternal and Child Care, Council in Medical Service, *A Guide for the Study of Perinatal Mortality and Morbidity*: 8.

population project that refashioned the care of pregnancy and childbirth, establishing novel conditions of existence for women's bodily experiences of having children.

Anthropologists and sociologists have demonstrated that the fetal imaging, screening and diagnosis found in our present have been used to constitute the fetus as a person/subject/individual, one that can be imagined as a member of varying collectivities such as nation, ethnicity and/or "race". The genealogy of the fetus at the perinatal threshold shows another mode of reasoning for the conservation of the fetus, reasoning that involved the alignment of a concept denoting a temporal interval (the perinatal) with a technique (risk assessment) for a governmental goal (the reduction of perinatal mortality). Rather than reading the public fetus as a reflex of proliferating fetal imagery, this account locates the maternal–fetal distinction in a temporal divergence: the split between the time of the perinatal and the time of pregnancy. The diagnostic testing that is part of enhancing the security of the fetus at the level of patient care links to the governance of the fetus at the level of national and international health policy. Standardized prenatal risk assessment acts as a higher order form of managerial reasoning in relation to ultrasound, amniocentesis, Rh-factor testing and so forth, incorporating these diagnostic tests into the framework of risk-based prenatal care.

"Perinatal outcomes", "perinatal care", "perinatal risk" pertain to the fetus/unborn and newborn. The subject of perinatal mortality is the fetus and newborn; it is a distinction within infant mortality, not within maternal mortality. After birth, the temporal distinction between perinatal and maternal is distributed over bodies separated at the end of the birth process – the mother and child. Prior to the separation of mother and child, the perinatal interval divides the time of the fetus/unborn from the time of pregnant women during pregnancy and labour. Yet the time of gestation *is* the time of pregnancy from the perspective of the previous threshold of the living subject, the birth threshold.

To compare the birth and perinatal thresholds is to juxtapose radically different historical experiences of embodiment that vary in relation to authorized knowledge, temporalities, and perceptions and conceptions of bodily substance. Unlike the birth threshold, which was focalized through women's embodied experience of pregnancy under conditions where practising physicians and midwives recognized they knew little about the unborn, and that little was dependent on the stories of women in their care, the perinatal threshold instantiates the perspectives of medical expertise and vital statistics.

The perinatal time span is dependent for its existence on two preexisting conditions: the standardization of the length of pregnancy as forty weeks and medical diagnosis of pregnancy. The twenty-eighth week of pregnancy is given its significance by being placed as an integer in a finite, numbered series: twenty-eight of forty, with twenty-eight reckoned as the lower limit of viability. Thus, the perinatal threshold presupposes the prior standardization of the length of pregnancy as forty weeks. This might seem an unremarkable observation, but the birth threshold had, from antiquity to the early modern period, been associated with a conception of pregnancy as having no fixed length, instead varying according to astrological conditions, the sex of the unborn, the heat of the womb and other conditions (Duden 2002a: 57–58; Duden 2002b: 14–15). The standardization of the temporal duration of pregnancy, begun in the *Elementa physiologiae corporis humani* (1766) of Albrecht von Haller (Duden 2002a: 64–65), imposed a Cartesian, linear time that acted to displace prior local temporalities, although this process has never been fully completed. Thus, for instance, "due date" in clinical practice is conventionally estimated from the time of "last menstrual period". The calculation of due date determines when a pregnancy is thought to have begun and when it should end. Notoriously, time of conception and time of last menstrual period may diverge significantly, and, although women in care may have accurate knowledge of time of conception, this knowledge is routinely ignored, with the consequence that due date is miscalculated and, as a result, women suffer unnecessary inductions of labour as they are perceived to have gone "beyond dates". The projection of a standardized length for pregnancy onto the bodies of pregnant women proceeds with a Cartesian linear simplicity that occurs at the expense of local temporalities, some of which continue to persist in the high modernity of the perinatal threshold.

The perinatal interval can exist only under conditions where pregnancy has become diagnosable, where the existence of pregnancy is determinable during its course, for otherwise prenatal risk assessment that includes "fetal factors" would be nonsensical. The first laboratory-based test for pregnancy was invented in 1928, and came into clinical practice during the 1930s (Oakley 1984: 96–98). The significance of these hormonal tests lay in making pregnancy a certain knowledge for both woman and health care provider during the time of pregnancy, prior to both the fetal heartbeat and quickening, in pointed contrast to the uncertainties of knowing pregnancy in the early modern period, which, even after quickening, became definitive only at birth. With the invention of hormonal tests for the diagnosis of pregnancy, birth no longer formed the moment of truth as to whether or not a woman was with child.

The birth and perinatal thresholds differ markedly in their conceptualization of who/what crosses the threshold of the living subject. At the perinatal threshold, the unborn is represented as a fetus that undergoes a fixed series of developmental stages; the mode of representation is realistic. The biological fetus is considered to exist independently of the mind, thus preexisting representation, but none the less capable of being perceived and known. In contrast, the early modern birth threshold represented the unborn through emblems, typically of boy children, which were not understood as realistic representations of the unborn but rather as symbolic of its presence (Duden 2002b: 18–30), that is as figurative rather than realistic representations of the unborn. At the birth threshold, the immanent and obscure presence of what Barbara Duden (2002a: 58–60) has called the *non-dum* – the not yet, hidden evidence of the unborn – was manifested as either a child or a "false fruit". The perinatal threshold clinically displaces the *non-dum* with the fetus during the last trimester of pregnancy and birth. At the perinatal threshold, the medical and statistical recognition of the living subject appears prior to birth, the threshold of medico-administrative perceptivity and reactivity lowered from the infant to the fetus during the last trimester of pregnancy and the birth process. Moreover, in contrast to the birth threshold, which finished at the conclusion of the birth process, the perinatal threshold continues after birth. The perinatal threshold has been extended into the hidden interior of pregnancy and beyond birth, a double extension that was done in the name of securing the health of national populations and children in clinical care.

The temporal organization of the perinatal threshold led to a problematization of the time at which a number of social statuses were to be conferred. Under the birth threshold, the *non-dum* was not considered a living subject, its status as such finalized only at birth. Once a child was born alive, a series of statuses was simultaneously and immediately conferred: that of person, individual, legal person, kin member, and, for increasing but still restricted numbers of living subjects after the French Revolution, citizen. The sharp distinction between the interior and exterior of a woman's body put into place through the birth threshold does not obtain at the perinatal threshold, where the living subject is present during pregnancy and birth. Prior to the last trimester, the unborn is imagined as an embryo and a fetus, but these have no governmental significance for the perinatal threshold; under its abstracted temporal schema death and illness exist solely from twenty-eight weeks gestation. The labour of conserving the health of population prior to and during birth destabilized the practice of assigning the statuses of person, individual, legal person, kin member, and citizen after birth. From the point at which the perinatal threshold was instituted, these statuses were potentially movable prior to birth and the simultaneity of their prior conferral dispersed. And so it came to pass that at the beginning of the twenty-first century it is possible, but not legally compulsory, for parents to grieve for babies born dead ("perinatal loss"), name them as individuals, identify them as sons and daughters, while at the same time it is settled at law in many jurisdictions that stillbirths have existence neither as legal persons nor as citizens. The displacement of the birth threshold by the perinatal threshold resulted in the problematization of the time at which ascribed statuses were to be assigned in relation to a living subject that transcended birth.

The perinatal threshold unsettles the relations of before and after, inside and outside held apart by the birth threshold. Living subjects at the perinatal threshold are located inside the bodies of women during late pregnancy and through the birth process. From the perspective of the birth threshold, the living subject only definitively exists after birth and outside the body of the woman in pregnancy and childbirth. The perinatal threshold brings the living subject into the time of pregnancy and childbirth and thus inside the body of another living subject. The biopolitics of the perinatal threshold acted to conserve the fetus/unborn for the sake of population health, but the status of that living subject for non-medical forms of social classification was unclear at the point when risk-based prenatal care was being constituted. Where the passage through the birth threshold simultaneously and indivisibly conferred the status of person and individual on the living subject, the effects of extending these statuses across the entire duration of the perinatal threshold raised questions pertaining to the liberal governance of women, questions unimagined by the physicians who had combined fetal and neonatal mortality for the purpose of enhancing security at the threshold of the living subject.

3 Health beyond risk

A midwifery ethos in prenatal care

The invention of prenatal risk assessment in the global north was linked to the governmental objective of reducing perinatal mortality rates. In Canada, risk-based prenatal assessment became standard practice for health care providers during the late 1970s and 1980s, with particular dates varying by province and territory. As I outlined in Chapter 2, a standardized prenatal risk assessment device designed by physicians in the Ontario Medical Association and distributed by the Ontario Ministry of Health first came into use in Ontario during 1980. Called the Ontario Antenatal Record (see appendix a, b and c to this chapter), it has remained in distribution with periodic updates since that time. "Antenatal Record": a record before birth. But whose record? Although it groups together a variety of factors – "maternal factors", "fetal factors", "social factors" – the Antenatal Record was devised to conserve population at the beginning of the perinatal interval. Its primary aim was to reduce perinatal mortality rather than maternal mortality, the latter being regarded as at an "irreducible minimum" when the Antenatal Record was first devised. The Antenatal Record concerted a meticulous clinical practice for the security of the living subject at the perinatal threshold.

The genealogy of standardized prenatal risk screening offered herein harmonizes with other social science accounts that have shown the extraordinary growth of risk governance in the latter half of the twentieth century and the beginning of the twenty-first. Social scientists have documented an avalanche of risk governance in this period, from the practices of welfare states insuring citizens against risk to the techniques of advanced liberalism constituting subjects as responsible for their own risk management. In this voluminous social science literature on the proliferation of risk judgement there has been little curiosity about possible objections to risk reasoning and their significance. There are, to be sure, exceptions to the general enthusiasm for documenting the proliferation of risk managerial reasoning. A variety of studies dealing with health and illness have examined how people in clinical care interpret information about risk given by physicians and other health care workers. This literature demonstrates that lay/everyday knowledge contains relevances and forms of reasoning that differ from the knowledge of expertise (Cox and McKellin 1999; Rapp 1989 and 1999). Other studies have focused on contestation between social movements and "experts", with expertise clothing itself in the mantle of "neutral" science (Michael 1992; Slovic 1987; Wynne 1995 and 1996). These lines of inquiry have richly documented the importance of popular knowledge as a normative resource for use against the risk techniques of expertise.

In the spirit of these studies the present chapter likewise investigates objections to risk reasoning, but the dispute lies *within expertise* rather than between lay and expert knowledge. I focus on the relation of expertise to risk-based prenatal care through a case study of midwives in the Canadian province of Ontario during the years following their attainment of legal recognition as a health care profession in 1994. Using interview and archival sources, I examine the relation between midwifery and risk-based prenatal care at two sites: clinical practice and legal "scope of practice" (the mandated actions a profession may perform under statutory law). The topic of risk in midwifery clinical practice was approached through interviews with midwives about how they orient to the task of completing the Ontario Antenatal Record. Only after midwifery became a legally recognized health care profession in Ontario at the beginning of 1994 were the majority of Ontario midwives first required to complete the Antenatal Record. The period of research was one in which Ontario midwives were actively negotiating their formal inclusion as a health care profession within a preexisting system of obstetrical care. Interaction with risk-based prenatal care thus formed a mark of midwifery's inclusion in the formal health care sector.

The use of interviews as a research source is methodologically anomalous for the history of the present, which has privileged archival sources and printed documents. My aim here does not consist of explicating a rationalized system of expertise – the focus of the history of the present – instead exploring how a health profession takes up a preexisting governmental technique that challenges aspects of its professional ethos. Where the previous chapter investigated the invention of prenatal risk assessment, the present one examines lines of fissure (if not flight) in risk governance, turning to a combination of oral and printed sources in order to access practitioner concerns and objections to risk governance that have not reached the monumental stability of the archive.

The focus of my work here is on the risk assessor's relation to risk, an area under-investigated in the social sciences with the important exception of Kemshall's (1998) work on risk assessment by social workers in probation practice. Kemshall shows that risk assessment is not a simple matter of applying risk based on probability calculations, but rather is a situated knowledge importantly mediated by "the value system of the risk assessor", values held in common with "the community of practitioners" (Kemshall 1998: 136). The present work echoes Kemshall's discussion, placing the relation of midwifery to risk-based prenatal care in the context of a professional ethos that seeks to change the culture of childbirth.

A number of studies, many undertaken by midwives themselves, have demonstrated that midwives possess a critical relation to risk. Saxell (1994) has shown the reluctance of midwives and their clients to classify pregnancy over

35 years of age as automatically a risk factor. Lane (1995), Murphy (1994) and Oakley and Houd (1990) have underscored the iatrogenic effects of risk governance in childbirth, arguing that risk screening routinely precipitates an increased rate of unnecessary interventions. Lane (1995: 65) has suggested that risk has "colonized debate" around childbirth, priorizing low perinatal mortality rates over any reference to patient experience in the evaluation of care. In alliance with these midwifery critiques, Oakley and Houd (1990) have argued that risk-based antenatal care mistakenly individualizes risk, which is better analysed in social and environmental terms, such as the concentration of risk in the underclass and lower working class due to poor nutrition, smoking, and low income. Tew (1990: 86–88 and 289), whose meticulous statistical studies have likewise been allied with midwifery in the critique of contemporary obstetrical methods, has maintained that the benefits of routine prenatal care are unproven except in the case of a small number of interventions, and even these proven interventions have had minimal impact on the reduction of maternal and child mortality. Tew primarily attributes the falling rate of maternal and child mortality in twentieth-century Britain (Tew 1990: 34 and 290) to maternal nutrition, during pregnancy and from the mother's own conception, a process cumulative over generations. The work presented here on the relation of Ontario midwives to prenatal risk assessment forms part of this research trajectory, confirming tensions between midwifery and risk-based prenatal care.

Internationally midwifery defines itself as the care of "physiologic"/ "healthy"/"normal"[1] pregnancy and birth. This self-definition commits midwifery to fashioning modes of care appropriate to pregnancy and birth in a state of health. It is simultaneously a declaration that the majority of pregnancies and births are "physiologic"/"healthy"/"normal" and should be managed as such rather than as "at no predictable risk"/"pathologic"/"unhealthy"/"abnormal". This orientation represents what I will call the ethos of midwifery, following Foucault's remarks in his discussion of the meaning of the European Enlightenment for our present. Foucault suggested that "ethos" be considered as an attitude: a way of thinking and feeling that "marks a relation of belonging and presents itself as a task" (Foucault 1997f: 309). His few comments on ethos are situated in a long history of its use in the Western rhetorical tradition, where ethos has been considered part of persuasive rhetoric. In Aristotle, ethos names the rhetorical efforts of a speaker attempting to persuade an audience on the basis of projecting a personal character which the speaker hopes the audience will find laudable. Aristotelian rhetoric distinguishes *ethos* from *pathos*, which appeals to emotion, and from *logos*, which appeals to reason (Corbett 1971: 93–99).

Another sense of ethos appears in Max Weber's sociology of religion, where Weber used ethos to conceptualize the attachment and commitment that persons feel to the groups of which they form part (Bendix 1977: 260–261). Weber's sense of ethos was closer to Foucault's "relation of belonging" that "presents itself as a task" than it was to ethos in Aristotle's *Rhetoric*. Foucault's treatment of the "ethos of Enlightenment" linked ethos to an ethical task, but also to the work of reason, thus forming an intrinsic relation between ethos and logos. One might note here that the genealogy of "ethics" in our contemporary sense is derived from the Ancient Greek *ethikos*, theory of living, which has as its root the most ancient meanings of ethos as place of living/custom. Discussions of ethos fall proximate to ethics and to ways of living. Foucault's critical revision of ethos requires its reconceptualization as an attitude that attempts to reconcile an ethical way of living with reason.

For the purpose of understanding midwifery in advanced modernity, I take ethos as a collective speaking/writing position, a reflexively elaborated group belonging developed in dialogue with research and practice. Ethos in this sense continuously projects a bridge across the modern dichotomy of fact and value. The contemporary midwifery ethos investigated here forms a normative relation between midwifery and women during pregnancy and childbirth. This commitment is normative, but its condition of existence is also the collective technical knowledge of expertise.

The understanding of pregnancy as a state of health is both a task and a way of belonging for midwifery, a profoundly normative claim. In advanced modernity the ethos of midwifery orients to methods of care that reduce intervention rates. In addition to the client's physiological needs, she is perceived to be emotional, social and cultural in ways relevant to clinical care. Pregnancy and birth are viewed as life transitions, with midwife as counsellor and guide. The goals of prenatal care for midwifery are thus broader than lowering perinatal mortality and morbidity, which is the aim of risk-based care.

The present study shows that Ontario midwives distinguish clinical from risk reasoning in prenatal care. Midwifery research participants argued against the application of risk judgements to women's bodies in clinical practice. Fundamentally, midwives objected to clinical opinions about the normal and the pathological being organized through risk reasoning. However, it would be incorrect to argue that midwifery has been entirely oppositional to risk-based prenatal care. More precisely, midwifery has privileged an older form of clinical reasoning over standardized risk assessment, thereby creating a complex field of tensions between clinical and risk reasoning.

The professional ethos of midwifery operates in the context of law regulating the health professions. When midwifery became a legally recognized health care profession in Ontario during 1994, its scope of practice was restricted by statute to "normal" pregnancy and birth, whereas the scope of practice for medicine was defined as including both "normal" and "abnormal". Midwifery's scope of practice thus legally required a means for distinguishing "normal" from "abnormal". Risk screening provided one possible method for dividing normal from abnormal, but the College of Midwives of Ontario opted to define the midwifery scope of practice in terms of "indications", a preference that privileged an older, clinical vocabulary over risk reasoning.

Recent discussions within the history of the present have emphasized the capacity of law to act as a mechanism of normalization (Ewald 1990; Hunt and

Wickham 1994: 49–52; Rose and Valverde 1998: 548–549) rather than as a power of prohibition. To effect normalization in health, law is articulated to discourses for the administration of life. The dependence of law on external expertise for specification of the normal opens law to intra- and interprofessional disputes about the parameters of the normal and techniques for its identification. As Rose and Valverde have observed (1998: 549), "Normalization does not describe an achievement, but rather a kind of mobile and heterogeneous transactional zone of conflict and alliance between forms of expertise." In the case of care during pregnancy and childbirth, the legal authorization of midwifery as a practice of "normal", and medicine as a practice of both "normal" and "abnormal", opened statutory law to agonistic fields of expertise and their frequently conflicting methods for producing the normal.

Not all normativity is normalizing. The "zone of conflict" between midwifery and medicine with respect to prenatal care is, I will argue, not strictly speaking one between normalizations but between normativity and normalization, a distinction based on differing styles of reasoning found in contemporary health care. While risk judgement conforms to a model of normalization, clinical reason does not: it is normative but not normalizing. Midwifery struggles to resist interpreting pregnancy and childbirth through risk reasoning, particularly in clinical practice. The tensions between clinical and risk reasoning in midwifery care illustrate plural styles of reasoning in contemporary health care.

Social scientific studies of contemporary Canadian midwifery have been framed in terms of the sociology of the professions, with the period since the mid-1980s theorized as a struggle for the formation of a new profession (see below for the recent history of midwifery in Ontario). This literature has argued that, once professional recognition was achieved, midwifery acted as a self-interested elite group with monopoly control (Benoit 1994; Bourgeault 1996; Shroff 1997). The postulated trajectory begins in a pure pre-professional moment, resolutely anti-medical, and ends with medical capture at professionalization – an historical reading implicated in the romance of the marginal to which the social sciences are sometimes prone. The present study of Ontario midwifery and its relation to risk techniques, being situated in the history of the present, offers a different understanding of expertise and the professions (Lupton 1997; Rose 1994a). The power of the professions lies not in taking away and hoarding and shutting down, but in implanting new desires, optimizing resources and inciting action – a positive rather than a negative model of power. As Rose notes, the expertise–subject relation is not one reducible to domination; expertise operates through "subjectification", "making up" persons who are stabilized in relation to a grid of norms and knowledges (Rose 1994a: 362). Professional knowledge operates, not through monopoly, but through the "generosity of expertise" (Rose 1994a: 361–362), lending knowledge and explanation to other social groups and occupations, although such generosity extends less strongly to property.

The understanding of expertise and its relation to knowledge found within the history of the present takes apart a simple binary of medical vs. midwifery knowledge caught in a relation of authority and domination. This way of writing unproductively stereotypes both medicine and midwifery. Medicine is not homogeneous, containing a single knowledge system that is authoritative over and exterior to midwifery. Medicine contains differences (Berg and Mol 1998: 1–12); it is an agonistic field of competitive positions and knowledge claims. Thus, the two main sources for prenatal care guidelines in Canada during the period of research, those of the Society of Obstetricians and Gynaecologists of Canada (1995) "Healthy Beginnings: Guidelines for Care during Pregnancy and Childbirth" and the Canadian Task Force on the Periodic Health Examination (1994), *The Canadian Guide to Clinical Preventative Health Care*, contain conflicting recommendations for prenatal risk assessment. Those of the Society of Obstetricians and Gynacologists of Canada are based on research and expert opinion, while those of the Canadian Task Force are research-based only. The *Guidelines* of the Society of Obstetricians and Gynecologists contain a much longer list of risk factors than those of the Task Force. This represents a difference *within* medicine. In discussions and debates about risk-based prenatal care and childbirth, Ontario midwives have consistently allied with research-based guidelines and critiques, many associated with evidence-based medicine. Midwives have not been anti-medicine, but rather have selectively accepted research-based medical knowledge to limit risk-based care.

History and ethos of Ontario midwifery

Some basic information about Canadian midwifery history is necessary in order to understand the context in which Ontario midwifery came to elaborate its relations to risk-based prenatal care. Throughout most of the twentieth century, Canada was the only Euroamerican industrialized country with no legal provision for the practice and profession of midwifery. Medicine claimed sole professional mandate in the provision of primary maternity care (for Ontario see Biggs 1983; Connor 1994; for an overview of Canada see Biggs 2004; *Report of the Task Force on the Implementation of Midwifery* 1987). Lay midwives continued to attend births in the decades prior to World War II, and the practice of midwifery was maintained among Aboriginal peoples, although undermined in the colonial encounter (Carroll and Benoit 2004: 269–271; Mitchinson 2002: 80–82). After World War II, the only exception to this pattern of physician-attended births, other than among Aboriginal peoples, occurred in remote nursing stations unserviced by physicians where a small number of trained (usually British) midwives were employed by the government (Benoit 1991; Mitchinson 2002: 72–75).

The 1970s and 1980s marked a period of international social movement activism that questioned existing medical practices related to pregnancy and childbirth.[2] This political mobilizing characterized maternity care as impersonal, sometimes degrading, and lacking respect for women and women's choices during birth (Mathews and Zadak 1991). Called the "home birth movement" in Ontario, organizations argued for the provision of home births and against the policy of universal hospital confinement during childbirth. The home birth movement was particularly critical of the medical institution, advocating lay midwifery as an alternative to physician care (Bourgeault 1996: 5–6; Burtch 1994: 103–134). Advocacy was accompanied by the growth of community midwifery assisting women in home births and acting as advocates for birthing women in hospitals (Tyson *et al.* 1995). Lay midwives formed groups to promote the legal recognition of midwifery and education for midwives, the main Ontario organization being the Association of Ontario Midwives. Social movement activists and lay midwives were galvanized by a key event during 1985: a lengthy coroner's inquest regarding the death of a baby at a midwife-attended birth (Van Wagner 1988: 116–117). The prosecution attempted to turn the inquest into a general condemnation of midwifery in favour of exclusively physician-attended births. The possible use of criminal sanctions against lay midwives[3] gave rise to a political response that led from a provincial task force to passage of the *Midwifery Act, 1991* (Bourgeault 2000; Bourgeault 1996: 59–92; Van Wagner 2004), which gave legal recognition to midwifery as a health care profession.[4]

Social movement advocacy for midwifery, home births and lower-intensity care during pregnancy and birth were directed against the risk-based system of prenatal care and birth that had been put into place with the rise of perinatal medicine during the 1960s. The fundamental question of what constituted appropriate pregnancy and birth technology was raised in response to the policy of universal hospital confinement during childbirth and universal risk-based prenatal care for lowering perinatal mortality and morbidity rates. The reduction of the perinatal mortality rate was linked to higher-level intensities of care for women; the social movement activism of the 1970s and 1980s problematized this policy, recommending mass non-compliance. The activism of the 1970s and 1980s took issue with the effects of risk-based perinatal care, representing a critical women's response to the profound historical transformation in the governance of pregnancy and childbirth that had taken place in an astonishingly short historical period from roughly 1930 to 1970.

A history of political advocacy, in conjunction with international discussions of community midwifery, defined the emergent professional ethos of midwifery during the 1990s when midwifery came to be legally recognized as an independent health profession in the Canadian provinces of Ontario, Alberta, British Columbia and Manitoba. The ethos was crystallized in a document produced by the College of Midwives of Ontario (the regulatory body constituted through the *Midwifery Act*): "Philosophy of Midwifery Care in Ontario" (College of Midwives of Ontario 1994a), a statement of fundamental principles (see Figure 3.1). The first of its principles states: "Midwifery care is based on a respect for pregnancy as a state of health and childbirth as a normal physiologic process and a profound event in a woman's life." The mode of intervention that follows from this principle commits the practitioner to a style of pregnancy and childbirth management that reduces intervention rates. Additionally it increases the counselling and educational component of care during pregnancy and birth. The

COLLEGE OF MIDWIVES OF ONTARIO
L'ORDRE DE SAGES-FEMMES DE L'ONTARIO

PHILOSOPHY OF MIDWIFERY CARE IN ONTARIO

Midwifery care is based on a respect for pregnancy as a state of health and childbirth as a normal physiologic process and a profound event in a woman's life.

Midwifery care respects the diversity of women's needs and the variety of personal and cultural meanings which women, family and communities bring to the pregnancy, birth and early parenting experience.

The maintenance and promotion of health throughout the childbearing cycle are central to midwifery care. Midwives focus on preventive care and the appropriate use of technology.

Care is continuous, personalized and non-authoritarian. It responds to a woman's social, emotional and cultural as well as physical needs.

Midwives respect the woman's right to choice of caregiver and place of birth in accordance with the Standards of Practice of the College of Midwives. Midwives are willing to attend birth in a variety of settings, including birth at home.

Midwives encourage the woman to actively participate in her care throughout pregnancy, birth and postpartum period and make choices about the manner in which her care is provided.

Midwifery care includes education and counseling, enabling a woman to make informed choices.

Midwives promote decision-making as a shared responsibility, between the woman, her family (as defined by the woman) and her caregivers. The mother is recognized as the primary decision maker.

Midwives regard the interests of the woman and the fetus as compatible. They focus their care on the mother to obtain the best outcomes for the woman and her newborn.

Fundamental to midwifery care is the understanding that a woman's caregivers respect and support her so that she may give birth safe, with power and dignity.

The *Philosophy of Midwifery Care* was approved by the College of Midwives of Ontario in January 1994, the first month that the legal recognition of midwives became effective in Ontario. This document may be understood as a statement of a midwifery ethos in the sense that it sets out a professional identification and task cfor midwives in relation to women during pregnancy and birth. This commitment is normative, but its condition of existence is also collective technical knowledge. The professional ethos of midwifery bridges fact and value.

Figure 3.1 Philosophy of midwifery care in Ontario

Source: College of Midwives of Ontario. Reproduced with the permission of the College of Midwives of Ontario.

"Philosophy of Midwifery Care in Ontario" defines midwifery as a practice of "preventive care and appropriate technology", one that "includes education and counselling, enabling a woman to make informed choices". Care is to be "continuous, personalized and non-authoritarian", in relation to "a woman's social, emotional and cultural as well as physical needs". Decisions regarding care are a "shared responsibility between the woman, her family (as defined by the woman) and her caregivers", with the woman in care the "primary decision-maker".

The ethos elaborated in the "Philosophy of Midwifery Care in Ontario" invokes a model of expertise–subject relations as egalitarian, educational, and affective. It seeks to govern women in their freedom and to place pregnancy and birth in the context of kin and emotional relations. This midwifery ethos orients pregnancy care to practices that are not measurable solely in terms of perinatal mortality and morbidity, which are the objectives of risk-based prenatal care. The "Philosophy of Midwifery Care in Ontario" implicates six "practice principles" meant to distinguish the routine practice of midwifery from medical care: enabling choice of birthplace, spending adequate time with the woman, facilitating informed choice, maintaining continuity of care with a small group of known caregivers,[5] developing a personalized, egalitarian relationship with the woman, and appropriate use of technology. Physicians by contrast do not offer home births as an option,[6] have much shorter mean appointment times (five to ten minutes vs. the midwifery community standard of forty-five minutes), and have significantly higher intervention rates than midwives. The "Philosophy of Midwifery Care" and its six "practice principles" comprise a network of distinctions dividing midwifery from medicine in the care of physiologic pregnancy and birth. These practice principles constitute midwifery's identifying features and profit of distinction.

The Antenatal Record: risk-based prenatal care

When Ontario midwifery became a legally recognized profession in January 1994 (the date the *Midwifery Act* became effective), midwives were for the first time entitled to apply for hospital privileges, among them the authority to admit clients to hospital under the care of a midwife. With the granting of admitting privileges came the obligation to comply with hospital protocols. Because all hospitals (save two) during 1994[7] stipulated that the Antenatal Record be submitted by health care providers for all women having in-hospital births, completing the Antenatal Record became a requirement for midwives for planned hospital births.[8] Hospital protocols ushered midwives into a preexisting system of risk-based prenatal care standardized in part through the Antenatal Record.[9]

The stipulation that midwives complete the Antenatal Record and participate in risk-based prenatal record keeping was a sign of their legally authorized acceptance in the formal health care sector. Physician use of the Antenatal Record preceded that of midwives. As mentioned above, since its inception in 1980, the

Antenatal Record had been distributed by the Ontario Ministry of Health in conjunction with the Ontario Medical Association. It was thus already part of the culture of medicine, and enjoyed high rates of physician support (Hanvey *et al.* 1991: 9–10).

The hospital requirement to submit a completed Antenatal Record for in-hospital births placed midwives in the position of transmitting risk management into clinical practice. Although judgements of risk and judgements of the normal and the pathological are, as I argued in the previous chapter, analytically distinct, in clinical practice both physicians and patients interpret risk as a sign of being between a state of health and illness, neither sick nor healthy. However, the ethos of midwifery takes a strong stand on pregnancy as a state of health. The stipulation to complete the Antenatal Record implicated midwives in risk-based prenatal care, a practice at variance with midwifery's stated ethos.

To understand the points of conflict between the ethos of midwifery and risk-based prenatal care, some acquaintance with the specifics of the Antenatal Record as a transmitter of standardized risk governance is needed. Although I am about to describe a particular document from the province of Ontario, Canada, similar standardized risk assessment texts are found wherever the reduction of perinatal mortality has been taken as a health care objective, that is, throughout the global north (Buekens 1995; Schücking and Schwarz 2002). If the account I provide is burdened with a certain microsociological zeal, it serves a compensatory function, showing descriptively how risk is relayed from vital statistics about population health to the bodies of individuals in clinical practice. This descriptive labour is intended as a corrective to accounts of risk that leave its relation to practice abstract, unclear, and subject to what Dorothy Smith has called "mystical connections" (Smith 1990: 45–51).

The first version of the Ontario Antenatal Record appeared in 1980 (see Chapter 2). The edition described and analysed herein (see appendix a, b and c to this chapter) was in use during the period of research (1995–1998); this version of the Antenatal Record was distributed by the Ontario Ministry of Health with minor revisions from 1992 until 2002.[10] The Antenatal Record is a complex, multifunctional document that acts as a record, a communication device, and an organizer of consultations and referrals across health care professionals. The 1992–2002 version consisted of two pages, with Antenatal Record 1 being the history and physical examination taken at the first visit, and Antenatal Record 2 the documentation taken at subsequent visits (appendix: 112–114). Each page has on its reverse side the "Guide to Pregnancy Risk Assessment": a list of risk factors graded into three levels of increasing seriousness (see appendix, p. 114).

The information elicited on Antenatal 1 and 2 included obstetrical, medical and family histories, the results of physical examinations and diagnostic tests, clinical signs and symptoms, plus notes for "social risk factors" and "discussion topics". Copies of Antenatal 1 and 2 were forwarded to the hospital for planned in-hospital births. In more complex clinical cases needing consultation and referrals beyond the primary health care provider (midwife or family physician), a copy was forwarded to the consultant (physician, obstetrician, perinatologist). The Antenatal Record operated, and its current revision continues to operate, as a communication device using a familiar format to link health care providers and deliver information to the place of birth (hospital or home), where it may be needed during childbirth.

The Antenatal Record organized referrals and place of birth by three risk levels associated with a tripartite division in intensity of care. One can see how this works by looking at the base of Antenatal Record 1 (appendix a: 112) where there is a place for "Risk Factors Identified" and, in bold, a box for risk assessment in terms of three options: A – "Healthy Pregnancy, no predictable risk", B – "Pregnancy at risk", and C – "Pregnancy at high risk". The three levels appear again in the "Guide to Pregnancy Risk Assessment" (appendix c: 114) on the reverse sides of Antenatal Record 1 and 2, with level A indicating the absence of risk while levels B and C correspond to the presence of one or more risk factors. Antenatal Record 2 (appendix b: 113) also has a column for risk assessment that prompts the assignment of a risk level at each prenatal visit. Risk levels A, B and C connect to differing levels of "intensity of care": those designated "A" remain with the primary health care provider (midwife or family physician), while "B" may require consultation with an obstetrician or internist and birth at a hospital with a secondary level of care. "C" indicates an emergency requiring immediate referral and preferable transfer to a specialized perinatal centre at the tertiary level. The three risk designations implicate a tertiary, regional model of health care.[11]

Clinical risk is marked by the combination of risk judgement with diagnostics/therapeutics. This is true of the Antenatal Record, where no clear demarcation between risk and diagnostics/therapy is discernible. Scanning the Antenatal Record, it is difficult to separate which categories pertain to risk and which do not. Diagnostic categories appear as risk factors in the "Guide to Pregnancy Risk Assessment" (appendix c: 114), e.g. "diabetes", "renal disease without hypertension", "heart disease, especially with failure", and "other significant medical illness". The presence of illness can thus confer prenatal risk. Clinical signs are also included as risk factors: "premature rupture of membranes" and "hydramnios", as are so-called lifestyle factors: "significant tobacco, alcohol, drug intake" and "severe social problems". Treatments are also included as risk factors: "rhesus immunization". Clearly, name, address and telephone number are unrelated to risk, but other demographic information such as date of birth (to establish the risk factor "primigravida (age 35 yr +)", and occupation may be relevant. In the elision between risk and diagnostics/therapeutics, risk factors proliferate and the state of health becomes elusive, a condition of "no predictable risk", that is, uncertainty. The "Guide to Pregnancy Risk Assessment" conflates risk with patient management, that is, any intervention by a health care provider. In this process, risk invades the space of clinical reasoning.

Because I was interested in the ways midwives reason about clinical risk, interview research was a more appropriate method than looking at completed Antenatal Records. In order to know whether or not the accounts provided by midwives were unique to midwifery or shared with other health care providers, the research design needed to be comparative. Between January 1995 and August 1998 I interviewed twenty midwives[12] and a comparison group of twenty family physicians,[13] who were similar to midwives in that they exclusively provided care for "low-risk" pregnancy and childbirth. All midwives interviewed had been in practice prior to the legal recognition of midwifery in 1994. Obstetricians were not included in the sample as they are specialists in pregnancies classified as "higher risk".

The interviews reflect the experience of practitioners at a moment of professional transformation as they were being incorporated into a state-funded health care system, all the while trying to maintain and formulate the distinctiveness of their midwifery practices. Having been undertaken at this historical moment, the research gives powerful insight into how midwives reason about risk. The study may also be of interest to midwives in reflecting on midwifery's complex relation to risk governance. The research design evidences practices of reasoning that act to restrain the spread of risk governance within midwifery, but it should be noted that this is not a generalizing methodology showing the extent to which midwives in Ontario, Canada or internationally use or do not use risk judgements in clinical care. However, midwifery clinical practice is required to conform to the standards of the College of Midwives of Ontario, and the practice standards of the College have legal force. Since, as I will show, the College practice standard is formulated in terms of "indications" rather than "risk" , midwives were not legally required to use the Guide to Risk Assessment or to perform risk assessments on women in care. The relation of midwives to the College practice standard is a legally binding, generalizing relation.

The interview format was semi-structured, with all questions open-ended. Participants were given or requested to have a copy of the Antenatal Record for reference during the interview. The interview elicited a detailed description of how the participant routinely filled in the form, with prompts asking about deletions, blanks, substitutions, additions and objections to existing categories. There were specific prompts for the boxes on "Family Situation", "Social Risk Factors" and "Risk Assessment" (both the "Risk Assessment" box at the base of Antenatal 1 and the column for risk assessment on Antenatal 2 (appendix a and b: 112–113)).

Over the course of the interviews, midwifery research participants referred to discussions about risk that had occurred in the early 1990s during the process of developing professional standards for the College of Midwives of Ontario. As a result, I interviewed three midwives who had been on the Standards and Professional Relations Committee of the Transitional Council of Midwives of Ontario (the body that preceded the establishment of College of Midwives of

Ontario).[14] Archived documents of the Standards and Professional Relations Committee held at the offices of the College of Midwives of Ontario were also reviewed.

Prior to legal recognition in January 1994, midwifery practice groups used a variety of internal recording forms with no standardization of contents across practices. The Association of Ontario Midwives had been developing and distributing suggested forms since the mid-1980s, but these were not uniformly accepted. The Antenatal Record did not initially displace these prior forms in all practices: nine of the twenty midwives interviewed reported double recording on their own forms and the Antenatal Record, gradually changing to the Antenatal Record due to, as one woman put it, "the reality of my time" (Midwife 12, 2 October 1996).

The Antenatal Record did not comprise the only type of documentation midwives and physicians kept on their clients. Physicians reported keeping what they called "progress notes" over and above the Antenatal Record. Midwifery research participants similarly reported using blank pages of paper called "narrative notes" on which they recorded information or observations in excess of the standard practice forms and the Antenatal Record. However, midwives used an additional series of forms that were circulated within practice groups rather than being shared with hospitals. These forms collected information on topics that were either absent from the Antenatal Record or treated in a less detailed fashion: client's pregnancy and birth history, family obstetrical history (especially of the woman's mother), a three- to five-day dietary history and a questionnaire for optional client use.[15] The questionnaire included queries about ideas, fears, religious beliefs, feelings about bodily changes, and partner's first and present reaction to her pregnancy. Midwifery practice forms consistently asked about pregnancy and childbirth experientially and emotionally. These topics signal an ethos of clinical care in excess of that found in the risk-based prenatal care systematized through the Antenatal Record.

Charting clinical risk

Hospital requirements for the submission of the Antenatal Record potentially placed midwives in the position of charting[16] clients in terms of risk. Midwifery research participants, however, thought that risk-based prenatal care undermined the understanding of pregnancy as a state of health.[17] For midwives, pregnancy care was governed by objectives broader than lowering the perinatal mortality and morbidity rate. In the interviews midwives identified three main areas of disagreement with respect to clinical risk in the Antenatal Record: the effects of assessing risk levels (A, B, C), the stigma associated with social risk factors, and the configuration of the expertise–subject relation. Their disagreements on this last point involved a related presence and absence in the Antenatal Record, which they saw as implanting a relation of non-collaborative authority between caregiver and client. Since midwives understood the subject in pregnancy as social and emotional, they took their own position in expertise to be that of educator and counsellor.

The concerns of midwifery participants were shared in part by physicians. Midwives and physicians had a common unease with recording "social risk factors", feeling that this invaded the privacy of their clients and stigmatized them in hospital. Physician participants wanted additions and deletions of particular risk factors, but with one exception did not share midwives' objections to risk assessment as such. Midwife and physician participants diverged with respect to perceived omissions from the Antenatal Record, with physicians wanting single category additions, whereas midwives wanted to include whole categories of absent information, information that did not necessarily pertain to reducing perinatal mortality and morbidity: the social, emotional and experiential aspects of pregnancy.

Pregnancy and risk levels

Few midwifery research participants reported regularly completing the risk assessment boxes on Antenatal Record 1 and 2 with a risk level: A – pregnancy at no predictable risk, B – pregnancy at risk, C – pregnancy at high risk. Thirteen of the twenty participants reported never having assigned a risk level, and of these, eight crossed out the risk assessment column for each prenatal visit, substituting "fetal movement". A further five reported sometimes filling in risk assessment, typically as A (no predictable risk), two doing so always, though noting their accountability to their College standard, "Indications for Mandatory Discussion, Consultation and Transfer of Care".

Midwifery research participants reported they did not use the "Guide to Pregnancy Risk Assessment". Their disengagement was emphatic:

> We do not use the Guide. We ignore this part of the form. We have our own separate methods for what you could call risk screening.
>
> (Midwife 20, 27 April 1995)

> Indications are pertinent to midwives, not risk categories.
>
> (Midwife 4, 12 December 1995)

> I just read this today.
>
> (Midwife 2, 24 June 1996)

> This is developed for physicians, this form, and even in terms of this categorization of risk, this isn't the categorization or even the process midwives use.
>
> (Midwife 1, 30 September 1996)

Midwives attributed their disregard of the "Guide to Pregnancy Risk Assessment" to a professional and legal obligation to comply with the practice standard of the College of Midwives, "Indications for Mandatory Discussion, Consultation and Transfer of Care". This document lists the conditions of abnormality or borderline abnormality in prenatal care and childbirth, grouping them together under three "categories" of increasing seriousness, the first requiring discussion with other practice partners or physicians sharing primary care, the second a consultation with a physician and the third a transfer to a physician for primary care. (The significance of the College standard is discussed below under "Scope of practice".) Tellingly, the College standard defined the boundary of midwifery in terms of "indications", a clinical concept, rather than "risk", with the result that midwives were not compelled to categorize the abnormal in pregnancy through risk, but through indications (discussed below).

Global criticisms of risk assessment were made both by midwives and, as I will later show, by physicians. Two midwifery participants contrasted risk assessment with recording clinical signs and symptoms:

> No, I don't fill it in, no. I can't stand the risk assessment … I can't stand these notions of urgency or emergency or drama that's in the word of risk, terrible, abnormal and tragic notion to the word and I can't stand it. I would like to see, for example – you know where they just have the remarks and list factors identified? – I want just lines. I want only comments or factors or whatever. It is important that I know this woman had a stillbirth. It's important to know she had a caesarian. It is important to me that she would have a fine, totally normal birth, and just because she is a grand multipara, she's not a risk. She's just having her sixth baby, and something like that I want to communicate factors that seem interesting … Is she an A, or is she a B, or is she C? Go home, if you want to list your alphabet … I already know what's the matter with the women, I don't want to know anything else and I don't want to categorize them for it. It's a punishment.
>
> (Midwife 9, 25 July 1996)

> The way I approach this is not as a risk assessment. I approach it as what's the story here? What's the picture? What are you dealing with through the giving of the information, but not necessarily to assess it for risk.
>
> LW: Would you explain that point to me?
>
> What's the story of your previous births. I mean, if there was a big bleed, it's noted. Or if there was difficulty or trouble, if it was a breech, if there was a caesarian section, it's noted. It becomes part of the story, and it's taken into consideration in terms of how I might approach the pregnancy and birth this time. On the other hand if somebody is coming to see me and they have diabetes, then I refer them to a physician for assessment.
>
> (Midwife 11, 23 September 1996)

In these passages, the two participants defended clinical reasoning and charting with clinical signs and symptoms against risk assessment (A, B, C) and charting with risk factors. The primacy and autonomy of clinical reasoning was maintained against risk assessments of women in care: "if there was a big bleed, it's noted"; "It is important to know this woman had a stillbirth." The research participants accepted that clients with diabetes should be referred to obstetricians under the "Indications for Mandatory Discussion, Consultation and Transfer of Care" as outside the scope of midwifery practice. But for those clients in midwifery care, risk assessment as "terrible, abnormal and tragic" would take apart the space of the "fine, totally normal birth". The sober flat lines needed for recording a client's history and clinical "story" deflate the hyperbole of risk.

Two midwifery research participants objected to *assessing* risk as "judging women", something they declined to do, preferring to "discuss" or record women's "stories", a mix of experiential and clinical recording. Midwifery participants differed from physician participants in valuing the role of "stories" within clinical care. Charting was itself referred to as recording a story. "Story" was used contrastive with "risk", which was typed as a form of reasoning that puts an end to a story.

One midwife who stated that she occasionally completed the risk assessment none the less objected to the extraordinary growth of risk discourse around contemporary pregnancy:

> The introduction of risk scoring into clinical practice carries the danger of replacing a potential risk of adverse outcome with the certain risk of dubious treatment and intervention … I am quite aware if somebody has a physical problem in pregnancy that requires, you know, an obstetrician … but I need to focus with women, my focus is on normal. More likely even if they're at risk or deemed at risk, the outcome will be good, right? And if I keep focusing on, well, you're more at risk, blah, blah, blah, then maybe I can make it happen … I find it quite disturbing actually that you have to talk so much about risk in pregnancy now. I think this is one of the great changes that has happened over the years that I've been going to births is this constant discussion of risk … I find it quite worrisome actually, that this is what the focus has become.
>
> (Midwife 17, 2 October 1996)

This midwife expressed concern about the constant orientation to risk in clinical care. She emphasized that practitioners treat clients for risk factors, leading to unnecessary interventions that cause ill health (iatrogenic effects). Risk, she argued, diverted clinical care from the "normal", precipitated the abnormal. She contextualized this concern in her opening comments, which are a close paraphrase of a critical article on risk scoring systems, Alexander and Keirse's (1989: 361) "Formal Risk Scoring during Pregnancy" in *Effective Care in Pregnancy and Childbirth.* Called ECPC (the letters are pronounced) informally by midwives, *Effective Care in Pregnancy and Childbirth* forms a central text of midwifery education internationally; it is situated in a research-based critique of obstetrics.

The above objections to risk assessment by midwives do not suffer from a simplistic view of pregnancy as a perpetually physiologic state never requiring intervention, a criticism often made by those who naively equate high-intensity care with safety. The midwifery research participants uniformly agreed with the need for recording clinical signs and symptoms, while simultaneously objecting to the effects of risk assessment. In any case, due to the College of Midwives practice standard, risk assessment was not relevant to prenatal care given by midwives. Risk, they argued, oriented practice to worst case scenarios and consequently precipitated increased levels of medical intervention, a strategy opposed to maximizing the place of "health" and the "normal" in the interpretation of pregnancy and childbirth. The hyperbole of risk contrasted with the sober plain style of charting clinical signs and symptoms.

Whereas eighteen of the twenty midwifery participants reported filling in the risk assessment boxes on Antenatal Record 1 and 2 never or only occasionally, eighteen of the physician participants stated that they generally completed them, with eleven acknowledging occasional omissions, particularly on Antenatal Record 2 when no change of status occurred from visit to visit. The physicians critically engaged with the "Guide to Pregnancy Risk Assessment", focusing exclusively on the appropriateness and inappropriateness of risk factors in Level B. The objections were twofold. Certain risk factors such as "maternal obesity" did not conform to the criteria of research-based medicine and either were questionable or should not be listed as risk factors; they were a "mish-mash, not predictively useful" (Physician 3, 11 December 1997). Second, some risk factors[18] did not constitute appropriate bases for referral or consultation, since obstetricians would handle the situation in the same way as family physicians. The participants (who were all family physicians) felt themselves more competent than obstetricians to deal with certain factors:

> I would be sending these women down to see the obstetrician for a consultation based on the fact that they're B pregnancy at risk, and their problems are that they don't have any food in the house, or they don't have a stable place to live, or their boyfriend beats them or there's a combination of factors like that. Well, what good is the obstetrician going to do for that, you know (laughter).
>
> (Physician 14, 22 October 1997)

> I'm probably better at handling severe social problems than an obstetrician is.
>
> (Physician 9, 15 December 1997)

Other physician participants found that consultants did not accept referrals for certain kinds of risk factors:

> Some things are overrated and not really considered by the people I refer them to as risk factors warranting referral, for example, primagravida age 35 years+. The consultants don't want to see those patients. Give you another example: cigarette smoking. You know, they don't want to see them. It's a risk factor but not a basis for referral … most of the people in my place, they don't want to know about breeches until they're in labour.
>
> (Physician 11, 16 December 1997)

These physicians objected to how particular risk factors gave rise to unnecessary referrals.[19]

Thirteen of the physician participants suggested additions to the Antenatal Record that would expand the number of risk factors, particularly HIV testing, STD screening for gonorrhea and chlamydia, infections, and the taking of folic acid. Fully five physicians objected to routine glucose screening, arguing that it increased intervention rates without benefiting patients. Eight physicians mentioned the need for more space to cover "psychosocial issues". The categories in the medical and family history sections of Antenatal Record 1 were questioned as not evidence-based – "a joke" as two physicians put it – with the prominence given to "early onset deafness", puzzling several sceptical physicians.

Global objections to risk assessment in the Antenatal Record were made by two physician participants; with the criticisms related to health funding and risk effects. One physician passionately disagreed with health care funding priorities, drawing an analogy between antenatal and neonatal risk screening:

> We're still spending a huge amount of money on identification and very little money on management and that goes through the whole health care system. And I guess it's just because it's easier that way; it's easier to make up forms like they've just done for the newborns. And, you know, have a grading system. I mean, that's the easy way out, to find, sure, you've just delivered a newborn to some 14-year-old drug addict with no family support and you've got a child at risk. Great. This child has a point score of fourteen when the cutoff is nine. So everyone's really pleased that we've now identified this kind of risk. And then what. You know? Public Health'll go out there once or twice. Children's Aid will say they're too busy, and there's no funding around to look after this kind, you know, and yet the government's now going to tout that we've got this wonderful new thing in place for identifying risk right at birth.
>
> (Physician 15, 25 June 1998)

This physician placed clinical uses of risk within a logic of "identification" that functioned badly for "social risk factors". Screening populations for risk may identify those having a risk factor, but acute care methods are effective neither for prevention nor for treatment of "social risk factors". From this practitioner's perspective, the result was completely useless for the purpose of referrals. One might note here that this physician was objecting to the ineffective rendering of an epidemiological norm as a disciplinary norm in clinical risk (see below for discussion).

A second global objection to risk assessment was voiced by a physician participant who was concerned about the negative effects of risk identification on women in care:

> It's philosophically questionable whether labelling the person high risk, which almost invariably within our system sets them up for this program of increased surveillance, without really improving the odds they're going to get a good outcome, but certainly making them self-identify themselves as being at risk, self-identify their babies as being at risk … I think it matters if someone is given the perception that their pregnancy is not normal. I think this matters enormously … I think it is most helpful for women to be told in a very positive tone that they fall within the range of normal. Not to be told in a negative tone that they're bigger than average or smaller than average. Which in our society is pejorative. Most physicians think this is a trivial point.
>
> (Physician 3, 11 December 1997)

This last point echoes comments of the midwifery participants regarding the emphasis of the Antenatal Record on abnormality, rather than on "pregnancy as positive and exciting" (Midwife 17, 2 October 1996). As one of the midwives put it with intentionally comic effect, "let's tell her what great haemoglobin she has" (Midwife 9, 25 July 1996).

The main line of criticism of the Antenatal Record among physician participants involved replacing individual risk factors with ones they judged more appropriate, criticisms that would amount to a new edition of the Antenatal Record rather than a questioning of governing pregnancy through risk. Two physicians had much broader critiques of prenatal risk governance, one objecting to identification without supportive action, and the other to the harm risk assessment causes women's self-perception. Since the Antenatal Record provided for clinical judgement and discretion in assessment, consultations and referrals, physicians did not feel their actions to be tightly constrained by it. Physician research participants established a critical, self-reflexive relation to the risk factors and standards of risk assessment found in the Antenatal Record, but did not problematize their relation to risk governance generally.

Social risk factors and stigma

The Antenatal Record is a component of a risk-based prenatal system governed by the objective of lowering perinatal mortality and morbidity and thus its "social" categories are subordinated to this end. Charting "social risk factors" presented challenges to both midwifery and physician participants, for each group regarded these factors as personally sensitive information about clients. Midwifery participants, and to a lesser extent physicians, were also concerned about charting social risk factors that might potentially stigmatize clients within hospital settings. Physician and midwifery participants entered into discussions with clients about whether and what to record, especially with respect to "Family situation", "Social risk factors", and to some extent even demographic information.

The potential readers of the Antenatal Record include clients, their partners and family as well as health care providers. This readership is in part due to the fact that a copy of the Antenatal Record is often given to women late in pregnancy so that it may be available at the place of birth, although in cases of planned hospital births, the copy may be forwarded directly to the hospital. From the perspective of health care providers in clinical practice, the addressee of the Antenatal Record includes other health care personnel, the client and potentially the client's kin.

Under Canadian law, all patients have a right to view their charts, but midwives are unusual in making a practice of giving photocopies to the client at the end of care. The College of Midwives of Ontario also requires as a matter of policy that its members share records with their clients, a policy linked to promoting client decision-making capacity:

> In the interest of promoting and maintaining informed choice, the midwife must make the midwifery care records available to the woman throughout her care. All midwives are obligated to provide a copy of the complete midwifery records to the woman upon request and within a reasonable amount of time following completion of care.[20]

During the period of research there existed no comparable requirement for physicians by the College of Physicians and Surgeons of Ontario.

The College of Midwives' policy on records stipulates that a client's chart be made available to her during care and a copy of the chart be given to her on request at the end of care. Local variation in implementing this policy was present among the midwifery participants, with some practices routinely copying the full file for every client and others giving a shorter summary of care. The practice of sharing the chart with a client was sometimes referred to as gift among a subgroup of midwives:

> There's a whole group of people who believe that notes are really about the midwife serving as a person who writes almost like a journal account of the woman – of what her experience is – and gives that to her as a gift.
>
> (Midwife 19, 19 June 1996)

Although none of the physician participants mentioned copying or giving abbreviated versions of charts to patients, seven did make a practice of giving the

Antenatal Record to the woman late in pregnancy, and another eight would give the woman a copy of her Antenatal Record when the physician was out of town. This was done to ensure the availability of the information at the place of birth.

Both midwives and physicians managed charting with an eye to how disclosure of information might damage client demeanour, anticipating that clients and their families might read the completed Antenatal Record, and, in the case of midwives, potentially the entire chart. Twelve of the midwifery participants stated that they talked with women in care at some point when completing Antenatal Record 1 about what they wanted recorded. They were concerned about client privacy and confidentiality, with the additional possibility of punitive care in hospital particularly in relation to "social risk factors". Physicians had similar concerns to midwives, with ten reporting that they discussed with their patients if/what to write down when completing one or more categories on Antenatal Record 1 (including Family situation and Social risk factors); these discussions involved information the physicians regarded as personally sensitive: unnecessary for health care, a violation of privacy or potentially revealing events the patient wanted concealed. Two physicians preferred to leave "Family situation" blank, orally communicating their observations in hospital.

The majority of physicians and midwives reported concerns that information written on the Antenatal Record would compromise client confidentiality in hospital and with the client's immediate family, and some midwives and physicians mentioned stigma in hospital. The confidentiality concerns were raised in the context of the "Family situation" and "Social risk factors" sections of the Antenatal Record. Two midwives reported leaving "Family situation" blank, three made it optional, and five completed it only positively, e.g. always writing "stable" when they chose to fill it in. When midwives and physicians were concerned about lack of support, the information would appear in narrative notes and consultation notes that stayed with the health care provider rather than being written on the Antenatal Record, which entered the woman's permanent chart in hospital.

Only one midwife and seven physicians routinely filled in "Social risk factors". Five midwives stated they never completed it, with another fourteen rarely doing so because they believed it stigmatized clients having hospital births or otherwise negatively affected care:

> This [the Antenatal Record] is the piece of paper that's going to the hospital when the woman is 36 weeks pregnant. If she sees any of the hospital staff, it's going to be for a very limited period of time. They are not going to be able to develop a relationship with you. They will judge her, and they do, on her physical appearance, how she presents in the hospital and what's on the piece of paper. And there can be profound decisions made "on her behalf".
>
> (Midwife 4, 12 December 1995)

> I rarely actually write something under there [Social risk factors] because to me it's flagged in a way that almost always looks punitive.
>
> (Midwife 19, 19 June 1996)

Midwifery participants placed sensitive information in narrative and consultation notes. In cases such as cocaine use that would make a difference to clinical care, midwives reported explaining to clients the necessity of sharing information in some fashion, sometimes with temporary cards or detachable notes, sometimes writing it on the Antenatal Record:

> If a woman's sharing something that seems really sensitive, I will say, "look, do you want me to make notes about this? Do you think you want me to share this with the other midwives caring for you? I won't write this in a permanent way but we'll need to know and discuss this so how about if I do I do it this way and I'll write a temporary card which gets thrown out".
>
> (Midwife 14, 21 June 1996)

Exactly what should be interpreted as a "social risk factor" was of concern to one midwife, who expressed a need for training in its interpretation. Reasoning that particular occupations had greater likelihood than others of exposure to communicable diseases or toxins, she was given pause to occasionally find herself writing "is a daycare worker" or "works with animals" as a "social risk factor" (Midwife 15, 9 May 1996).

Fully seven physicians spoke with patients about what to record under "Social risk factors", motivated by confidentiality and the importance of patient control over personal information:

> All medical records are supposed to be confidential, but of course, we all know that confidentiality doesn't often mean an awful lot. Things get faxed back and forth. I don't know that I think they might do anything bad with it. It's more, I suppose, that I'm just concerned that the patient should have some option of what she wants to disclose to various people who are involved in her care. So that she might choose to disclose to me, for example, a history of abuse when she was a child, or she might choose to disclose that to the social worker. She wouldn't necessarily choose to show that with six members of the nursing staff at the Hôtel Dieu[21] in the antenatal clinic, and another six when she goes into labour, you know?
>
> (Physician 14, 22 October 1997)

Two physicians purposely left "Social risk factors" blank; three recorded euphemisms and moved the information into progress notes, the contents of which were relayed verbally to hospital. Another physician said the need for recording "Social risk factors" did not arise in his practice.

The majority of both midwifery and physician research participants reported having discussions with women in care about recording information at a number of points in addition to "Family situation" and "Social risk factors". Therapeutic abortions unknown to partners, previous births ending in adoption and, particularly for midwives, "education level" and "marital status", precipitated collaborative discussions. Both groups repeatedly used the phrase "none of our business" for education level and marital status, commenting that this information was not relevant to clinical care.

The process of implementing a process of totalized risk screening in pregnancy eventuated in a series of subterfuges on the part of health care providers: textual strategies to record information separately from the Antenatal Record, tactics of oral communication to restrict circulation of information, euphemisms recorded on the form, and spaces left blank by design. Both midwives and physicians raised "privacy" concerns in relation to charting risk and were reluctant to disclose information that might be stigmatizing for women in care. The use of non-disclosure and euphemism for the "social" categories of the Antenatal Record comprised tactics for managing precarious information that might damage patient demeanour in hospital and among her kin.

Expertise–subject relations

The Antenatal Record prescribes a particular expertise–subject relation, with the provider positioned as questioner, the woman in care as respondent.[22] The questioner generates the topics of the interaction and is positioned as having the authority to make these selections under the wide warrant of health, ranging from occupation, contraception and family situation to protein sources and thyroid disease.[23] Midwives had systematic disagreements with the implied physician–patient relation presupposed by the Antenatal Record with respect to who they took themselves to be, whom they acted on, how they were to govern themselves, and to what ends they practised.

In "The Philosophy of Midwifery Care in Ontario" (see Figure 3.1), the College of Midwives projected a set of normative criteria for expertise–subject relations. Midwifery care is prescribed as "continuous, personalized, and non-authoritarian". Midwives are subjectified as educators and counsellors, with clients interpellated as social, emotional, and cultural "as well as" physical. This statement of professional ethos defines expertise as a collaborative achievement between caregiver and client. The expertise–subject relation in midwifery care positions the client as a locus of significant decision-making, which is modelled as "shared responsibility between the woman, her family (as defined by the woman and her caregivers), with the woman as primary". Shared decision-making is linked to the practice of "informed choice": the presentation of alternatives to clients for their decisions. A key point in the provision of informed choice by midwives relates to birthplace. Where the Antenatal Record assumed the hospital as the sole place of birth, midwives are required to give women in care a choice of birthplace, with home birth conceptualized as a powerful way of facilitating the entry of newborns into the life of families.

The normative practices endorsed in the "Philosophy of Midwifery Care in Ontario" exceed the implicit value placed on positive outcomes found in the Antenatal Record. Pregnancy in the Antenatal Record appears as a physiological event with a residue of "psychosocial risk factors" linked to perinatal mortality and morbidity. The information solicited in the "History of Pregnancy" for current pregnancy is, by way of example, rigorously limited to the pathophysiologic: bleeding, vomiting, pyrexia (fever), smoking, alcohol, and radiation.

Midwifery participants found the social, experiential and emotional aspects of pregnancy and childbirth notably absent from the Antenatal Record:

> Emotions don't exist on this form. I ask people about emotions and I will make notes where it's relevant.
>
> (Midwife 14, 21 June 1996)

> We're trying to help the woman deal with the pregnancy in more than just a physiological way … You have this tiny little space here [for history of current pregnancy]. So, have they bled? Have they vomited? Pyrexia? Smoking? Alcohol? Radiation? There's a thousand things that can happen in pregnancy that someone might be talking about. They might be talking about ambivalence about the pregnancy and you know, emotional changes. Once again, these may be things that are very important in terms of physiologic safety.
>
> (Midwife 20, 27 April 1995)

> When I would get care transferred to me, from a physician, it was more, more the act that there, the kinds of issues of how is a woman feeling about her pregnancy, were missing.
>
> (Midwife 10, 9 January 1998)

> That [the Antenatal Record] leaves out the global nature of the woman, the way she thinks and feels. It leaves out any kind of intuitive response we have with each other. It leaves out the family, the way they think.
>
> (Midwife 3, 7 October 1996)

Midwives viewed the Antenatal Record as framing pregnancy solely within pathophysiologic terms, contrasting this way of reasoning with the midwifery practice of considering pregnancy in emotional, social and familial terms. Information about a client's mother's obstetrical history may, the midwifery participants acknowledged, have little predictive value for outcomes measured in terms of perinatal mortality and morbidity, but may none the less be salient to a woman's feelings about her pregnancy, and thereby conducive to supportive prenatal care.

Midwifery research participants repeatedly remarked on the significance of narrative for their record-keeping, noting its absence from the Antenatal Record:

> The third page that doesn't exist here is the narrative. Where do I write in: moved to a new house, very stressed out, sister died last week or the things that are relevant to her care, flesh it out, that give a whole picture but is not assessing the woman?
>
> (Midwife 11, 23 September 1996)

> We're used to a more narrative kind of style documenting actual visits, when you write a bit about the visit and what you discussed and what your recommendations were, and what the woman's concerns were and what were her questions for that week.
>
> (Midwife 13, 9 July 1996)

> You know, midwifery notes tend to be a bit longwinded and they tend to reflect the fact that you care a lot about social issues as well and you care a lot about individualized stuff, so the notes may look a bit different from one person to another.
>
> (Midwife 19, 19 June 1996)

Note that the reaction of the midwifery research participants to the Antenatal Record did *not* consist of rejecting physiologic charting; their objection was to its exclusion of narrative record-keeping and the client's experiential, emotional and relational concerns.

A total of fourteen midwifery research participants referred to completing more than one copy of Antenatal Record 2 and/or mentioned having problems with insufficient space. Four physicians mentioned being constrained by space requirements. This divergence may reflect differences in the length of appointments, and the purposes of appointments, with the community standard for midwifery being forty-five minutes per appointment for all visits, whereas the practice for physicians was forty-five minutes for an initial visit, and five to ten minutes for succeeding visits. One physician participant raised concerns about the type of interaction generated by the Antenatal Record, finding it to be a "framework that drives the meetings", and a "questionnaire-style exchange" (Physician 18, 30 January 1995). The concerns about the interaction pattern put in place by the Antenatal Record were also shared by three midwifery participants.

One midwifery practice encouraged clients to take the position of expertise. A copy of Antenatal Record 2 was posted in the washroom with the heading "Filling Out Your Own Antenatal Record" and clients were asked to complete the date, weight, gestational age, and urine protein/glucose on their own charts. The authority of expertise was thus cheerfully inverted, and the expertise–subject relation modelled on shared responsibility.

To summarize the relation of midwifery to charting risk, one could say that, for midwives, prenatal care necessarily integrated social (including familial), cultural and emotional relevances that were missing from risk-based prenatal care, which was oriented to reducing perinatal mortality and morbidity. Midwives regarded preventing death and illness as essential but insufficient bases for prenatal care. Nor did they categorize pregnancy in terms of risk levels (A, B, C) as they were responsible to their College standard which used the category of "indications", not risk. Rather than charting through risk factors and risk assessments, they preferred to record clinical signs and symptoms, viewing this practice as consistent with their orientation to pregnancy as a state of health. The care of pregnancy in a state of health required a configuration of the care provider as educator and counsellor able to respond to client-generated topics. This expertise–subject relation complemented the subjectification of the client as a locus of shared decision-making and informed choice, including client control of information that might be stigmatizing in hospital. It also contrasted with the implied expertise–subject relation of the Antenatal Record, where the provider, but not the patient, knows what constitutes risk and even the discussion topics are provider-generated.

Midwives aimed not only to limit the place of risk in clinical care, but to expand care beyond the limit of risk. They brought relevances to care that were absent from risk-based prenatal care: advocacy for the interpretation of pregnancy as a state of health, and the stylization of the expertise–subject relation as collaborative. They were deeply concerned with the client as a "ritual self" (Goffman 1997: 27–35) stabilized through deference practices permitting her to maintain face with intimates and hospital personnel. Where risk-based prenatal care is formally unconcerned with emotions and social relations other than those rendered problematic as "psychosocial risk factors", midwives included a broader range of clients' feelings and intimate relations as significant to care, including those not deemed risk factors. Where the Antenatal Record read prenatal care in relation to risks pertinent to perinatal morbidity and mortality, midwives inscribed pregnancy and birth within kin solidarities that clinical care was to facilitate.

One cannot of course be so naive as to think that the constitution of the client as relevantly social and emotional for the purpose of care stands in some way outside power/knowledge relations. The analysis undertaken here has not taken the classic shape of what Thomas Osborne (1994) has called "anti-medicine" – a critique that writes the history of medicine as an authoritarian rejection of individuality and subjectivity. Following Foucault, Osborne decisively shows that clinical reason has from its inception in the late eighteenth century been a medicine of the individual body, and contemporary medicine, far from being insensitive to the subject, has developed an "expertise of subjectivity" (Osborne 1994: 42). In risk-based prenatal care, the "expertise of subjectivity" has been marginalized, a situation which has met with midwifery's expanded notion of prenatal care as significantly social, cultural and emotional. The situation has also been of concern to physicians. During the late 1990s the Subcommittee on the Antenatal Record of the Ontario Medical Association began work on an addition to the Antenatal Record for screening "psychosocial risk factors". The shadowy figure of the social and the emotional in risk-based obstetrics deserves further study.

Scope of practice: clinical indications vs. risk factors

When the College of Midwives of Ontario called pregnancy a "state of health" and childbirth a "normal physiologic process" ("Philosophy of Midwifery Care in Ontario" 1994a – see Figure 3.1), it gave a specific interpretation to the midwifery scope of practice found in the *Midwifery Act.* That *Act* restricted the midwifery scope of practice to the "normal" in pregnancy and childbirth:

> The practice of midwifery is the assessment and monitoring of women during pregnancy, labour and the post-partum period and of their newborn babies, the provision of care during normal pregnancy, labour and the post-partum period and the conducting of spontaneous normal vaginal deliveries.[24]

However, as Elisabeth Massey (1993: 351) has demonstrated, while the midwifery scope of practice at law is defined as the "normal", the *Medicine Act, 1991*[25] gives physicians the "exclusive authority to diagnose abnormal conditions in pregnancy and childbirth". Thus physicians have the authority to define the abnormal,[26] and thereby to determine how the normal and the abnormal will be divided. The abnormal specifies the legal boundary of midwifery and is thus constitutive of it. Midwives, for their part, are expected to recognize the abnormal in pregnancy and birth, referring such clients to physicians for consultation and potential transfer of care. The division between normal/abnormal in pregnancy and birth comprises a power/knowledge relation; the status group that constitutes this division controls the organization of maternal and child health. Until roughly the nineteenth century in Euroamerican jurisdictions, midwifery controlled that division, turning birth over to barber-surgeons when they understood labour as abnormal (Arney 1982: 25).

The division between the "normal" and the "abnormal" in pregnancy and childbirth (to use the words of the *Midwifery Act*) is neither singular nor stable in the contemporary health complex. Risk assessment is one method for accomplishing that division, and it is this method that takes institutional form in the Antenatal Record. Risk assessment divides pregnancy by risk levels and allocates the resulting divisions to differing professional groups, with midwifery delegated low-risk pregnancies and births. Midwifery, however, does not constitute the division between the normal and the abnormal through risk, but through clinical reasoning. The boundary between the "normal" and the "abnormal" is legally defined through the practice standard of their College, the "Indications for Mandatory Discussion, Consultation and Transfer of Care". The presence of "indications" rather than "risk" in this standard disarticulates Ontario midwifery practice from interpreting pregnancy and childbirth through risk.

Midwifery research participants suffered a conflict of jurisdictions with respect to the Antenatal Record. Compelled to complete it due to hospital protocols, their College standard for scope of practice defined abnormality in terms of "indications" rather than risk, and thus they were not legally responsible to the referral network found in the "Guide to Pregnancy Risk Assessment", nor to its definition of healthy pregnancy as a condition of "no predictable risk". Their College standard formed a legal barrier to the risk classifications that have been associated with perinatal medicine.

Physician research participants were likewise bound by hospital protocols to submit the Antenatal Record, but since the form had been the work of their own professional association, the Ontario Medical Association,[27] no interprofessional jurisdictional conflict existed for them. The Antenatal Record defined the standard of care for physicians, as one physician put it, "Once it's there, it becomes a standard of care, so that if you don't fill in that box [risk assessment] there's an omission from your care" (Physician 15, 25 June 1998).[28]

In the practice standard, "Indications for Mandatory Discussion, Consultation and Transfer of Care", the College of Midwives defined the midwifery scope of practice in an older, clinical vocabulary related to treatment rather than to the newer vocabulary of risk. This practice standard refers to "indications" throughout rather than "risk factors". "Indications" is a term of art found in therapeutics: "A sign or circumstance that indicates the proper treatment of a disease" (*Taber's Cyclopedic Medical Dictionary* 1997).[29] *Dorland's Medical Dictionary* (2003) defines indication as: "a sign or circumstance which points to or shows the cause, pathology, treatment or issue of an attack of disease; that which points out; that which serves as a guide or warning". *Stedman's Medical Dictionary* is terser, an indication being "the basis for using a particular treatment or diagnostic test". An indication may be based on knowledge of cause (causal indication), it may be a symptomatic knowledge without a known cause (symptomatic indication), or it may be a course of action that arises from what is understood to be the nature of a disease (specific indication). The presence of an indication points to an available intervention that will benefit the patient. To use "indications" in the context of discussion, consultation and transfer of care would appear to distance this concept from its proximity to therapeutics; the meaning of indication has shifted in the midwifery practice standard.

Preference for "indications" over "risk" as a means of distinguishing midwifery from medicine was not a consensus position in Ontario midwifery prior to its statutory recognition as a profession. During the late 1980s the Association of Ontario Midwives had formulated its own "Guidelines to Scope of Practice" in terms of "risk factors" rather than "indications". Engaged in developing standards for midwifery prior to legislation, as early as 1989 the Association of Ontario Midwives[30] had devised what it called a "Risk Screening/Consultation Protocol". This Protocol grouped together three levels of "risk factors" as a guide to consultation and transfer of care among providers.[31] In contrast, the form for charting women in care that was produced and distributed by the Association of Ontario Midwives did not use "risk" anywhere. The Association defined scope of practice in terms of risk, but did not recommend charting risk, that is, applying risk categories to women in care.

The initial drafts of the "Indications for Mandatory Discussion, Consultation and Transfer of Care" used "risk factors",[32] which was replaced in late 1993 by "indications".[33] The change reflected discussion in committee. One former member of the Committee on Standards and Professional Relations, which had been charged with drafting the criteria for consultations and transfer of care, recalled that "risk" seemed to "signify medical intervention", with "indications" being a "less inflammatory language" more suited to bearing in mind that "a pregnancy can still be normal and healthy while being classified as having a 'risk factor' " (Midwife 21, 19 June 1998). Another member of the Committee on Standards and Professional Relations recalled a debate about whether "indications" or "risk" should be used, with "indications" selected on the grounds that it priorized clinical judgement. As one former member of that Committee remarked:

> We wanted to see the woman as, you know, someone who's healthy, even if she has some of these indications, still may have, in many ways, a very normal healthy experience, and that there are clinical indications that may lead to risk. But we don't want to label these as risk factors. And again, there was sort of a critique of that as time went on. That [risk] began to get removed as the document evolved.
>
> (Midwife 1, 30 September 1996)

The relation between "indications" and "risk" appears strained in this passage, with the distinction defended as significant, but its analytic significance obscure.

A physician's comment made during the consultations of the Committee on Standards and Professional Relations with the Ontario Medical Association,[34] prompted the Committee to delete risk factors from the drafting of what became the "Indications for Mandatory Discussion, Consultation and Transfer of Care":

> In fact, when we were developing this and drafting it, we went to meet with the College of Physicians and Surgeons, and with the OMA [Ontario Medical Association], and it was actually one physician, I think it was at the OMA, who said, "You're using the word risk! Why are midwives using the word risk?" They had removed some of that language from theirs and why were we using it? It was a bit of an embarrassment … yeah, but the old document hadn't quite got cleaned up properly, so, you know.
>
> (Midwife 1, 30 September 1996)

Thus, the College of Midwives of Ontario ultimately opted for "indications" in preference to "risk" to determine the midwifery scope of practice, although previously the same Standards and Professional Relations Committee and the Association of Ontario Midwives had defined scope of practice in terms of risk.

In the interviews, some midwifery research participants continued to use "risk" in defining their scope of practice:

> Our particular practice sends out something in the mail that explains a little bit about our practice. It's two pages, and then another two pages which is a questionnaire … So when we get that back, if she's high risk, then we say, "Sorry, we can't take you on".
>
> (Midwife 6, 26 September 1996)

> During the initial discussion that you have with her before you even put pen to paper, you will have assessed her risk-wise you know, because she would have told you a bit about her previous pregnancies or about her health in general and if she's already telling you that she's a diabetic that takes insulin we can't really take her on as a client.
>
> (Midwife 5, 24 June 1996)

One midwife insisted on the importance of institutionalizing midwifery as a practice of normal birth so as to create a new culture of birth. She argued that this goal necessitated a form of screening procedure to divide normal from abnormal pregnancies and labours, but was completely neutral as to whether the division was to be made under the rubric of risk or indications (Midwife 19, 21 June 1996).

Ontario midwives are legally accountable to the definition of their scope of practice found in the "Indications for Mandatory Discussion, Consultation and Transfer of Care", together with whatever written modifications of this document individual practices may formally enter into with particular hospitals where practice members have admitting privileges. Thus, for the purposes of clinical care, the "Guide to Pregancy Risk Assessment" with its structure of referrals defined through risk factors and the risk assessment boxes on Antenatal Record 1 and 2 represented a form of classification not relevant to midwifery. This is illustrated in a case reported by one of the midwifery research participants. Caught in a conflict between her own assessment made according to the "Indications for Mandatory Discussion, Consultation and Transfer of Care", and a physician assessment made according to the "Guide to Pregnancy Risk Assessment", this midwife was told by the College of Midwives that she was obligated to follow the standard of her College and retain client care.

Risk governance was thus not new to the post-1994 period of midwifery professionalization, having been used by the Association of Ontario Midwives in the late 1980s and early 1990s to define scope of practice. Initial discussions in the Standards and Professional Relations Committee continued to formulate scope of practice in terms of risk, with "indications" later displacing risk on the grounds that it privileged clinical judgement. The meaning of the "normal" in the *Midwifery Act* thus came to be defined through "indications" rather than "risk" by the College of Midwives. The College practice standard, "Indications for Mandatory Discussion, Consultation and Transfer of Care", which has legal force, acts as an interpretive barrier to the spread of risk governance through midwifery, conceptualizing the abnormal in relation to indications rather than risk, and disarticulating risk from clinical practice. When, post-1994, midwives were granted admitting privileges to hospitals, they were bound by hospital protocols to submit a copy of the Ontario Antenatal Record for each in-hospital birth. The result was a conflict between a standard of the College of Midwives, "Indications for Mandatory Discussion, Consultation and Transfer of Care", and the "Guide to Pregnancy Risk Assessment" in the Antenatal Record. Statutory law granted midwifery the practice of "normal" maternity care, but the processes for identifying "normal" are stipulated by the health professions themselves. Provincial law was thus normalizing with respect to health practices during pregnancy and birth, but the pragmatics of normal were subject to dispute between medicine and midwifery.

Clinical reason and risk: normativity beyond normalization

In this chapter I have raised the general issue of critiques of and limits to risk governance found in our present. The capacity of midwifery to have a critical relation to risk-based prenatal care arose in the context of an ethos forged in an interprofessional dispute with medicine regarding the forms of clinical care appropriate to physiologic pregnancy and childbirth. In this dispute midwifery privileged an older, pre-risk form of clinical discourse for defining the normal both in its scope of professional practice and in practice methods. The tactic of orienting to the normal/healthy/physiologic through the use of clinical reasoning has acted to constrain the spread of risk governance in midwifery. At a more general level, this study points to the significance of styles of reasoning that are irreducible to risk and which act to break the proliferation of risk governance in our present.

The relation between clinical and risk reasoning in midwifery is much more complex than a line of opposition. A core–periphery metaphor of their relation would be more apt. Midwives sometimes, both in the period prior to professional recognition and in the period of research, defined the boundaries of their practice, including consultations and transfers of care to physicians, in terms of risk, while being resistant to using risk categories in clinical care. Health and the normal in clinical practice were not interpreted through risk judgements. Risk-based prenatal care was judged unsuitable for care of physiologic pregnancy because it resulted in unnecessary interventions with adverse effects. It was also thought to be too narrow; midwifery care of pregnancy in a state of health included documentation of the client as social, emotional and cultural. The midwifery expertise–subject relation foregrounded collaborative exchange, subordinating the reporting of stigmatizing "risk factors" to the needs of the client as a ritual self, requiring the maintenance of face. The governance of pregnancy as a state of health had for midwives an ethical basis that was sometimes at variance with risk-based prenatal care. Midwifery clinical practice for pregnancy in a state of health contained statements and practices exceeding the risk factors employed to reduce perinatal mortality and morbidity.

Midwifery does not have a relation of pure exteriority with respect to risk-based prenatal care. Because risk judgements saturate prenatal care in medicine, midwives must use risk as a *lingua franca* when addressing the medical institution. The present research has shown how the College standard, "Indications for Mandatory Discussion, Consultation and Transfer of Care", in its initial drafts defined scope of practice through risk factors, later shifting to indications. The history of the College standard shows that midwifery's relation to risk governance has in part been characterized by ambivalence, that is, a contradictory condition of alternating attachment and detachment. In light of this ambivalence, clarifying the analytic distinction between the judgement of risk and the judgement of the normal and the pathological in clinical reason might prove helpful. The distinction between risk and health/illness is an epistemic one that exists prior to research-based evaluations of prenatal risk assessment, which have as their condition the concept of risk.

During the last decades of the twentieth century and the early years of the twenty-first century midwifery defended pregnancy as a state of health and midwifery as the care of "physiological", "healthy", "normal" pregnancy and birth. The midwifery commitment to treat pregnancy as a state of health is analytically distinct from, and in tension with, the interpretation of pregnancy as a state of risk. As discussed in Chapters 1 and 2, a risk factor predicts the likelihood of *future* illnesses, while diagnosis of the normal and the pathological concerns *present* health/illness. When a person is diagnosed as being in a state of health or illness, the judgement is made on the basis of signs (including test results) and symptoms found in that patient's body. The transposition of risk from the epidemiological analysis of populations into clinical practice leads to the reconstruction of risk factors as clinical signs and symptoms located in a patient's body, with the consequence that people are treated for risk factors (Gifford 1986; Adelswärd and Sachs 1998: 200).

Formally, bodies at risk are neither healthy nor ill. The judgement of risk and the judgement of the normal and the pathological operate independently, and thus nothing about a person's health status can be known from the presence of a risk factor. However, in clinical practice both health care providers and patients interpret risk factors as a sign of being *between* a state of health and a state of illness (Gifford 1986). In their study of the clinical recontextualization of epidemiological risk, Adelswärd and Sachs (1998: 205) argue that, in clinical risk, "[t]he borderlines between risk of disease, disease and worry over disease become fuzzy". Clinical risk inserts a third term between health and illness that destabilizes their relation, rendering it "ambiguous" (ibid.). In contrast, the ethos of midwifery takes a strong normative and scientific stand on pregnancy as a state of health, not illness, not a state *between* health and illness. At stake in the relation between midwifery and risk is a professional ethos dedicated to creating a new birth culture.

Midwives are well aware that pregnancies in a state of health may develop "complications"/conditions requiring interventions by health care providers. The transition from midwifery care of pregnancy in a state of health to what midwives refer to as "abnormal pregnancy and birth" is organized through the concept of "indications" in the Ontario case, a concept learned from the historical collaboration of Ontario with Dutch midwifery. An indication is not a diagnosis, but a situation where available evidence shows the patient will be better off with an intervention than without it. As prenatal care in our present has come to be characterized by a collapsing of patient management (any intervention by a health care provider) into risk such that even test results and diagnoses are designated as risk factors, those interventions that base themselves on indications constitute an epistemic space for thinking health care provision outside the capacious parameters of risk. If indication forms a concept primarily belonging to therapeutics, then its application to scope of practice extends the concept in ways that require future thought. Undervalued in contemporary health care, the concept of indication requires more robust thinking if it is to serve as a means of limiting the saturation of contemporary pregnancy with risk judgements.

Midwifery's attempts to delink risk reasoning from clinical practice are pertinent to the general question of law and normalization that has been opened up through the work of François Ewald (1990). Ewald has drawn attention to a common misreading of Foucault's *Discipline and Punish* as arguing that law acts purely negatively: a power of prohibition. A closer reading shows that Foucault maintained that law has acted as a mechanism of normalization for roughly the last two centuries: the "penality of the norm" (Foucault 1979: 183). Building on Ewald's exposition, Rose and Valverde (1998: 548) have called attention to the joint action of legal and non-legal knowledges, emphasizing the connection of law to apparatuses of governance in effecting normalization. Law, they point out, authorizes the activities and regulates the powers of those who engage in producing normalization. Intrinsically unstable, normalization comprises a "heterogeneous transactional zone of conflict between forms of expertise" (Rose and Valverde 1998: 549). Applying this line of argument to midwifery, one can say that statutory law authorizes midwifery to be a practice of assisting "normal" pregnancy and childbirth, but it does not define the normal other than to indicate that the normal birth is "spontaneous" (i.e. without induction), "vaginal", and to authorize the performance of specific acts to assist birth. Further specification of the "normal" is legally allocated to the professions of midwifery and medicine. This process of devolution opens law to authorizing many modes of normalization.

Midwifery is thus charged with a scope of practice that statute law defines as "normal" pregnancy and birth. But does this legal language authorize only modes of normalization? Is the care of "normal"/physiologic pregnancy legally confined to "normalization"? Statutory law allocates the power to further elaborate the "normal" and the "abnormal" to expertise, but does the administration of life proceed to divide normal from abnormal solely according to normalization? These questions turn on the meaning of "normalization" and whether it is identical to normativity. I argue that the health care professions contain ways of dividing the normal and the pathological that are normative but not normalizing: normativity beyond normalization. This position is intended both to save the technical meaning of normalization and to make a case for attending to the heterogeneity of styles of reasoning found inside the health complex.

What then is meant by that tender and overused word, normalization? Two main explications of normalization appear in Foucault's writings, the first in *Discipline and Punish* (1979 [1975]: 177–184), and the second, qualifying discussion in *"Society Must be Defended": Lectures at the Collège de France, 1975–1976* ((2003: 242–253); 1997: 215–225). In *Discipline and Punish,* Foucault treats normalizing judgement solely in relation to disciplinary power. Discipline acts on the bodies of individuals to optimize their utility through a regime of visibility, documentation and examination that renders each and all present and accounted for. Foucault's discussion in "*Society Must Be Defended*" extends his discussion of the norm into the biopolitics of populations, but this latter work strongly emphasizes that biopolitics constitutes its object quite differently than does discipline. Simply put, biopolitics is exercised over populations, while discipline operates over individual bodies. The security mechanisms of biopolitics (what I called "population power" in Chapter 1 to distinguish it from the multiple powers at play in biopolitics more generally) operate on general processes characterizing human life conceived biologically. Access to the species life of humanity occurs through action on general, measurable, abstractions, with Foucault (2003: 246; 1997b: 219) giving as examples lowering mortality rates, raising life expectancies, and increasing the birth rate. Whereas disciplinary mechanisms maximize utility at the level of individual bodies, security mechanisms regularize random events at the level of population: "security mechanisms have to be installed around the random event inherent in a population of living beings so as to optimize a state of life" (Foucault 2003: 246; 1997b: 219).

Both the disciplining of the individual body and the regulation of the life processes of populations employ the logic of the norm for their effectivity. The norms of discipline and regulation each operate by means of quantitative distributions. A norm is established by fixing the parameters of a field's immanent characteristics in order to plot a distribution that may be flexibly divided into a central tendency and tail ends on either side of the centre. Foucault (2003: 253; 1997: 225) argues that the norm links the differing "levels" of disciplinary and biopolitical power: "The norm is something that can be applied to both a body one wishes to discipline and a population one wishes to regularize." He assures readers that the regulatory norms which format biopolitical populations are pervasively articulated to disciplinary norms, an insight that needs further work.[35]

How then are we to understand the relation between normalization, perinatal mortality, prenatal risk assessment and clinical reason? Perinatal mortality operates as an abstraction pertaining to the aleatory processes characterizing the life of the human species considered biologically. The perinatal mortality *rate* standardizes perinatal mortality, enabling the identification of mortality above and below the standard rate for a given region, and the formation of longitudinal series showing historical variations. In Foucault's terms, the perinatal mortality rate functions as a regulatory norm operating at a population level rather than at the level of individual bodies.

In clinical risk, the regulatory norms of population power are projected to the level of individual bodies. Clinical risk attempts to bridge the gap between epidemiological calculation at the population level and the application of risk at the individual level, but this cannot be done formally as the likelihood of a positive or negative outcome for an individual from an exposure cannot be computed (see Chapter 1 for discussion). Regulatory norms that apply at a population level as risk estimates enter technical uncertainty at the level of the individual. Clinical risk spreads horizontally to invade clinical reason in order to increase reliability, becoming in this process an accretion of incommensurable forms of reasoning called risk (see Chapter 2). Prenatal risk assessment, an instance of clinical risk, aspires to disciplinary normalization, achieving it through an act of will under conditions of technical uncertainty in order to implement the value of health over each and all at the perinatal threshold.

Clinical reason does not conform to regulatory or disciplinary normalization, being an amalgamation of the two, an impossible power/knowledge relation. In clinical reason, pathology is conceptualized as a process immanent in the individual body, taking specific courses and expression in individual cases. The clinical gaze is individualizing, or, more accurately, the clinical gaze sees the general at play in its individual manifestations. Clinical reason is normative: through its processes of reasoning, opinions about health and illness are made. But it is not normalizing as it does not have the same technical form as normalizing judgement.

The division between the normal and the pathological in medicine is an ancient one and has historically taken many different forms not assimilable to the model of normalization. Canguilhem's *The Normal and the Pathological*, which has been (indirectly) fundamental to the history of the present, was in part a polemic against identifying the norm in physiology with the statistical average (Canguilhem 1989: 151–180). Canguilhem's book defended a Darwinian understanding of normativity as a practice of adaptation; biological life, Canguilhem argued, is not characterized by fixed statistical norms but by the capacity for adaptation, the generating of new norms. *The Normal and the Pathological* was an extended argument for distinguishing the normative processes proper to life from normalization, although one might note that the nature/culture distinction was stabler at the time of its original publication in 1943 than now. Specific modes of dividing the normal and the pathological are of less significance here than the general point that the relation between the two terms varies historically and has been subject to internal dispute.

What significance does the distinction between normativity and normalization have for understanding the articulation of law with expertise in the administration of life? The meaning of "normal" remained essentially incomplete in the *Midwifery Act* and the *Medicine Act* other than to specify its opposite as the "abnormal". Statutory law allocated the power to specify the relation between normal and abnormal to professional groups. Professional groups engaging in the administration of life use a variety of styles of reasoning to distinguish the normal from the pathological, only some of which are normalizing, but all of which are, like clinical reason, normative in their aim of optimizing health.

The concept of normativity beyond normalization is significant here because it points to styles of reasoning within the health complex that surpass risk and that act to limit and contain risk reasoning. Clinical reason acts to contain the proliferation of risk. The contestation of risk, indeed alternatives to risk reasoning, is enabled by the conception of normativity beyond normalization.

Appendix

The following three pages are a reproduction of the Ontario Antenatal Record in the version used between 1992 and 2002, the period of the research. Devised by the Ontario Medical Association, it was distributed by the Ontario Ministry of Health. The first page (Antenatal Record 1) is for charting a woman's medical history. The second (Antenatal Record 2) comprises a record of the current pregnancy. On the reverse sides of Antenatal Records 1 and 2 appeared the "Guide to Pregnancy Risk Assessment", which is the third page here.

Appendix a: Antenatal Record 1

Ministry of Health
Ontario

In conjunction with the Ontario Medical Association

Antenatal Record 1

Hospital/Centre

Woman's Name

Address

Phone # Home

Date of Birth	Age	Education level	Marital Status M CL S

Occupation | Phone # Work

Partner's Name | Age | Occupation

Birth Attendants	Midwife	Family Physician	Obs. Care	Baby Care	Health Number

Pregnancy Summary

Menstrual History
LMP (date) Certain Yes No
Cycle Duration

Contraception
Type Duration
Last used

Family Situation

History of Pregnancy Yes/No
Bleeding
Vomiting
Pyrexia
Smoking (cig./day)
Alcohol
Radiation

GRAVIDA	TERM	PREM	AB.	LIVE	MULTIPREG	E.D.C. By Date By Ultra sound	Revised E.D.C.

Obstetrical History

Number	Year	Sex	Gestation Age (wks)	Birth Wt. (Kg)	Duration of Labour	Place of birth	Type of Delivery	Comments: Re – Pregnancy, Perinatal Death, Abnormality, etc.

Medical Histories

Significant Medical History	Yes/No	Comments
Kidney Disease		
Heart Disease		
Hypertension		
Diabetes		
Infections		
Thyroid Disease		
Transfusions		
Operations		
Herpes		
Other		

Family History	Yes/No	Comments
Hypertension		
Diabetes		
Heart Disease		
Multiple Births		
Malformation		
Genetic Disease		
Early Onset Deafness		
Other		

Social Risk Factors

Initial Physical Examination

Height	Pre–Pregnancy Weight	Present Weight	B.P.

Check (✓) if normal

Head, Neck, Thyroid	☐	Vagina	☐
ENT	☐	Cervix	☐
Teeth	☐	Uterus	☐
Chest	☐	Adnexa	☐
Breasts	☐	Extremities	☐
Nipples	☐	Varicosities	☐
Heart Vasc. System	☐	Neurological	☐
Abdomen	☐	Cytology	☐
Vulva	☐		

Pelvic Architecture

Remarks (detail abnormal findings)

Risk Factors Identified

NUTRITION ASSESSMENT
Milk Products
Breads and Cereals
Fruits and Vegetables
Protein Sources

DISCUSSION TOPICS
Smoking
Alcohol
Drug Use
Work Plans
Exercise
Coitus
Pregnancy Compl.
Prenatal Classes
Dental Care
Breast Feeding

Age ≥ 35 years: Genetic Testing: Declined Yes ☐ No ☐ Offered Yes ☐ No ☐

RISK ASSESSMENT

Attendant's Signature | **Date this Attendance**

0374-64 (92/10) Canary – Mother's Chart– Forward to Hospital Pink – Attendant's copy White – Infant's chart 7530-4654

Appendix b: Antenatal Record 2

Ministry of Health Ontario

In conjunction with the Ontario Medical Association

Antenatal Record 2

Woman's Name

Address

Birth Attendants

RISK FACTORS, ALLERGIES, MEDICATIONS. Identified at Initial Visit

E.D.C. byDate

Revised E.D.C.

G | T | P | A | L

Pre-Preg Weight		Rubella Immune		HBs Ag		VDRL		Hb		Blood Group		RH	+	–	RH Antibodies		RH Gamma 28 weeks Given	

Subsequent visits

Date	WT. (lb./K).	G-Age Wk.	S-F HT.	Presn Posn.	FH	Urine Pr	Urine Gl	B.P.	Risk Ass.	Comments: (other lab. medications etc.)

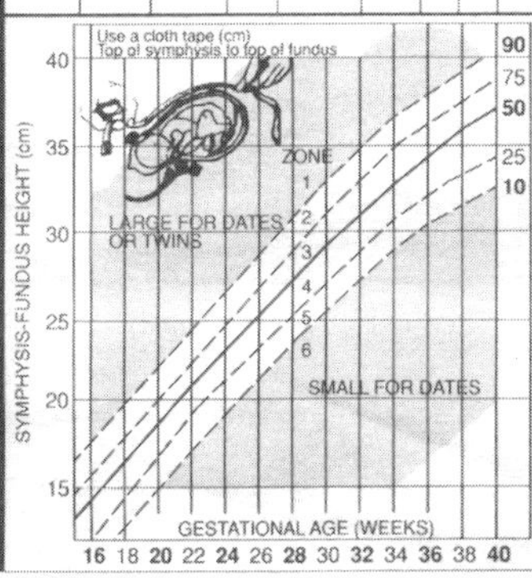

Ultrasound and Abnormal Laboratory Results

REFERRAL PLAN

- Public Health Nurse
- Dietitian
- Social Worker

DISCUSSION TOPICS

- Nutrition
- Diet
- Rest
- Travel
- Prem. Labour
- Breast Care
- Fetal Movement Counts
- V.B.A.C.
- Birth Plan
- Pain Relief
- Type of Delivery
- Labour Support
- Breast Feeding
- Room–in
- Circumcision
- Early Discharge
- P.P.S.P.
- Car Safety

Optional Investigations:

Alpha Feto Prot | T2 Ultra Sound | Glu. Screen

Additional Comments:

Measure symphysis-fundus height at each antenatal visit and record on graph opposite

Attendant's Signature:

0375–64 (92/12) 7530–4655

Appendix c: A Guide to Pregnancy Risk Assessment

A Guide to Pregnancy Risk Assessment

In the event of maternal transfer, please photocopy the front sheet and send to referral hospital

At each antenatal visit please give your assessment of pregnancy (fetal plus maternal) risk according to the following system. The risk factors or problems listed below are intended as examples only. Additional space is provided for other risk producing problems which you have identified. This assessment system is intended as a basis for planning the on–going management of the pregnancy.

A. Healthy Pregnancy, no predictable risk:

- ❑ No prior perinatal mortality or low birthweight infant
- ❑ No significant medical disease
- ❑ No pregnancy complications now or in the past (bleeding, hypertension, premature labour)
- ❑ Fetal growth seems adequate

B. Pregnancy at risk:

The fetus/mother may be at risk and consultation should be considered with a specialist obstetrician in your area. In addition, consultation with an appropriate internist may also be necessary. These patients may be managed by continuing collaborative care and birth in an obstetrical unit with intermediate level nursing facilities OR they may be returned to the care of the referring physician with a suggested plan of management for the remainder of the pregnancy.

- ❑ Diabetes, Class A (gestational) or Class B
- ❑ Hypertension without toxaemia
- ❑ APH, ceased and in hospital
- ❑ Cervical incompetence
- ❑ Hydramnios
- ❑ Post–date pregnancy (42 weeks +)
- ❑ History of prior stillbirth or neonatal death or premature delivery
- ❑ Maternal obesity (20%> ideal b.wt.)
- ❑ Significant tobacco, alcohol, drug intake
- ❑ Rhesus immunization
- ❑ Premature rupture of membranes 33 wk. or more
- ❑ Other significant medical illness
- ❑ Renal disease without hypertension
- ❑ Mild toxaemia
- ❑ Controlled premature labour
- ❑ Twin pregnancy
- ❑ Breech: beyond 36 weeks
- ❑ Primigravida (age 35 yr. +)
- ❑ History of genetic disease in family (Genetic amniocentesis or counselling required)
- ❑ Anaemic not responding to iron (Hgb <10gm)
- ❑ Weight gain <10lbs. by 30 weeks
- ❑ Grand multipara (>5)
- ❑ Large for dates
- ❑ Previous uterine surgery excluding lower segment caesarean section
- ❑ Severe social problems

C. Pregnancy at high risk:

Pregnancies which are so complicated that the fetus and/or mother are obviously in danger. If at all possible, these patients should be transferred to a regional perinatal centre (level III) for intensive care and birth. Clearly, there are patients who deserve to be placed in this risk category (with problems such as excessive antepartum bleeding, cord prolapse, or advanced uncontrolled premature labour) who cannot be transferred safely or in time to benefit the fetus or mother.

- ❑ Diabetes class C,D,F,R or significantly complicated
- ❑ Renal disease with hypertension ± ↓ function
- ❑ Premature rupture of membranes (± sepsis)*
- ❑ Antepartum bleeding, continuing or repeated*
- ❑ Fetal anomaly
- ❑ Multi–fetal gestation, more than two
- ❑ ______________________
- ❑ Hypertension with superimposed toxaemia
- ❑ Early uncontrolled premature labour*
- ❑ Severe fetal growth arrest (<10th percentile)
- ❑ Heart disease, especially with failure
- ❑ Oligo hydramnios
- ❑ ______________________

* 24–32 weeks gestational age

> ***Two or more*** risk problems can combine to produce a high pregnancy risk. Such a patient may need to be placed in a higher risk category

4 Legal fiction and reality effects

Evidence of perinatal risk

> [A]lthough the child was not actually born at the time the company by its fault created the conditions which brought about the deformity to its feet, yet, under the civil law, it is deemed to be so if for its advantage. Therefore when it was subsequently born alive and viable it was clothed with all the rights of action which it would have had if actually in existence at the date of the accident. The wrongful act of the company produced its damage on the birth of the child and the right of action was then complete.
>
> (*Montreal Tramways v. Léveillé* [1933], 4 D.L.R. at 344, Supreme Court of Canada)

> "There is no use trying," said Alice; "one can't believe impossible things." "I dare say you haven't had much practice," said the Queen. "When I was your age, I always did it for half an hour a day."
>
> (Lewis Carroll, *Alice in Wonderland*)

When the perinatal threshold of the living subject entered law during the last decades of the twentieth century, a relentless problematization of the legal subject and the legal address of pregnancy ensued. This chapter and the next deal with the legal reception of the perinatal threshold, with the present chapter focused on the legal concept, the child *en ventre sa mère*, and the next on child welfare litigation.

Until roughly the second half of the twentieth century in Canadian and other Euroamerican legal regimes, legal personhood had historically been defined by the birth threshold, with the partial exception of criminal law for offences after quickening. The legal person (other than corporate persons) had as its condition of existence the living subject recognized upon birth. In contrast, the perinatal threshold recognized temporal and bodily continuity between newborn and fetus. At law the perinatal threshold precipitated a range of new questions and possibilities, since its bodily and temporal continuities attached a division between woman and unborn (as "fetus") to the time of pregnancy and childbirth. In consequence, courts in Euroamerican jurisdictions repeatedly dealt with the question of what, if any, place law ought to have in conserving fetal health and life, a question that opened immediately to instituting a potential division in pregnancy between two legal persons: the unborn child and the woman carrying it. Because legal persons are rights-bearing, a problematization of the relation between mother and the unborn configured as fetus followed, posed in terms of conflicting rights. The stakes in these cases were many, notably the liberty rights of women, but also the intelligibility of the prior legal address of pregnancy, one aspect of which will be considered here: the child *en ventre sa mère*.[1]

An enormous scholarly and popular literature has examined attempts found in many legal regimes of the late twentieth and early twenty-first centuries to declare the fetus a legal person. In the midst of the highly politicized "fetal personhood" adjudication which directly challenged the liberties of women, the centuries-old legal address of pregnancy through the concept of the child *en ventre sa mère* has attracted much less notice, known to few other than a handful of legal scholars and feminists who were directly involved in the litigation that attempted to convert the child *en ventre sa mère* to a fetus in a *tour de force* of scientific reductionism.

The child *en ventre sa mère* is a *fictio legis* – a legal fiction – first formulated in ancient Roman inheritance law. "Fiction" carries a specific legal meaning in this context: a principle of legal reasoning that marks a concept as a nonliteral, figurative departure from ordinary usage. The fiction of the child *en ventre sa mère* permitted a child born alive, but still *in utero* at the time of a parent's death, to inherit property under conditions where the child had not been explicitly mentioned in the parent's will. Estate law treated the child *as if* s/he had been born at the time of the parent's death: a temporal fiction. This retrospective fiction applied only if the child were born alive (called the "live-birth rule"). Differing legal regimes reasoned that the intent of the deceased would have been to recognize and provide for the child after birth, and the fiction, called in Latin *qui in utero est*, continues to be used in both Anglophone common and Francophone civil law regimes into the present.[2]

From a concept that enabled the transmission of property and social status across generations, the child *en ventre sa mère* was transformed into a device for the bodily security of the living subject in twentieth-century tort law.[3] I argue that new nineteenth- and twentieth-century applications of the child *en ventre sa mère* concept were driven by risk reasoning, initially the actuarial risk of insurance, and then the epidemiological and clinical risk of the health complex. Beginning in the mid-nineteenth century, new applications of the child *en ventre sa mère* fiction paralleled the growth of the insurance industry, enabling children born alive after parental (primarily paternal) death in both workplace and motor vehicle accidents to receive survivor benefits. While inheritance law had not historically been implicated in risk techniques, the meshing of civil litigation with insurance thoroughly imbricated the child *en ventre sa mère* fiction in actuarial risk. Within insurance, risk is calculated actuarially by estimating the probability and amount of future losses over time, and then redistributing the anticipated loss across the members of the risk pool in order to protect them against the eventuality of sudden and large losses (Ewald 1991: 201–205). The new applications of the child *en ventre sa mère* were thus *actuarial effects* driven insurantially.[4]

The actuarial effects of insurance were conjoined with clinical and epidemiological risk when a new application of the child *en ventre sa mère* was created in twentieth-century tort law: the recognition of a right of action to recover damages for prenatal injuries. Suits for prenatal injuries have a consistent *topos*: a pregnant woman injured in a motor vehicle accident gives birth to a child with congenital health problems. A suit is launched in the child's name seeking financial compensation from the person who injured the mother. Claims for damages in suits for prenatal injuries were and are dependent on the existence of insurance, but also on biomedical evidence showing that harm to a child's health can be linked to conduct that injured fetal health: a risk judgement that crosses the birth threshold. Medical evidence about health effects across the birth threshold acted as a relay for the perinatal threshold from the health complex to the legal one. However, by application of the fiction, *en ventre sa mère*, the injury to the child is legally held to "crystallize" only after birth.

A 1933 judgement of the Supreme Court of Canada, *Montreal Tramways v. Léveillé,*[5] was the first decision internationally to allow medical evidence as proof of fact in claims involving prenatal injuries. Estimating the effects of an "accident" on the fetus linked prenatal injury with postnatal health problems across the birth threshold; without medical evidence of clinical/epidemiological risk, there could be no proof of prenatal injuries. Medical evidence in claims for prenatal injuries took the form of risk estimates: an opinion regarding the probability that an injury during pregnancy had caused the ill-health of a child later born alive. Physicians and other health care professionals reasoned from a particular child's health after birth to events that had occurred during pregnancy, using knowledge that arose from clinical experience and research literatures.

Recent writing in the history of the present has emphasized the interaction between governing through risk and governing through uncertainty (Ewald 2002; Haggerty 2003; O'Malley 2003 and 2004). As Janice Richardson (2004: 104) has noted, O'Malley's writings have been distinctive in examining the operation of risk and uncertainty within litigation and legal reasoning rather than externally at the level of insurance. In the case law dealing with claims for prenatal injuries examined here, both risk and uncertainty were at play. The legal subject was configured as acting under conditions of uncertainty, while expert evidence took the form of risk judgements that entered into legal reasoning in assessing liability. When twentieth-century tort law came to configure the legal subject as able to foresee the results of his/her actions over greater temporal and spatial spans than had been the case in previous tort law, and owing duties to more classes of persons, the fetus appeared on that expanded horizon as reasonably forseeable. In tort law, reasonable foreseeability norms the legal subject as having a duty to act to avoid loss and harm to others under conditions of uncertainty. Damage claims for prenatal injuries also mobilized medical estimates of risk. During the last two decades of the twentieth century, medical evidence linking fetal health with child health took the form of clinical risk estimates projected forwards and backwards across birth. When negligence was alleged, risk linked outcomes to exposures before, during and after birth, doing so under conditions that were technically uncertain, as physicians who gave evidence often remarked. Lapses of the legal subject in self-governance through uncertainty were thus evaluated through risk evidence that knew itself as uncertain.

The *fictio legis* of the child *en ventre sa mère* and the biological entity known as the fetus are radically heterogeneous concepts with differing histories. Although it is not presently known what form the threshold of the living subject took in Ancient Rome, the earliest applications of the child *en ventre sa mère* fiction considered herein date to early modern European history, where a birth threshold was the norm at law and in common culture. During the early modern period, birth was understood to make manifest what had previously been hidden in the depths of women's bodies, showing whether the unborn was a "true fruit" or another type of flesh. Duden (1999; 2002b) has clearly demonstrated that the concept of a fetus going through successive biological stages of development dates to Soemmerring's 1799 publication, *Icones Embryonum Humanorum*: an anatomical work, not a legal text. Thus from the early modern period until the early nineteenth century, the fiction of the child *en ventre sa mère* existed without reference to the biological fetus, securing the property rights and lineage of children *in utero* at the time of the parental death. Under the historical conditions of the birth threshold, only a live birth demonstrated that a woman had been pregnant. The child *en ventre sa mère* was not a biological entity; unlike the fetus, it could not have a heartbeat.

It was only with the 1933 decision in *Montreal Tramways* that the child *en ventre sa mère* and the biological fetus first came to be juxtaposed. When the Court admitted expert medical testimony in a suit involving alleged prenatal injuries, where previously the admission of such evidence had been denied, a realistic, referential, biomedical concept of the fetus was introduced into legal decisions employing the child *en ventre sa mère* fiction. The resulting concatenation of fetus with fiction gave rise to pervasive *reality effects* on the fiction of the child *en ventre sa mère,* destabilizing it *qua* fiction. Over several decades in the latter half of the twentieth century, the tensions between *fictio* and fetus threatened to dissolve the legal framework supporting the fiction of the child *en ventre sa mère* in order to secure the health of the fetus for the sake of children's health after birth. In this process the perinatal threshold came into competition with the birth threshold, attempting a fundamental revision of tort law principles. As we will see in this chapter and the next, the Supreme Court of Canada responded to the problem of legal intelligibility by reaffirming a distinction between legal and medical discourse, upholding the prior legal framework of tort law inclusive of a right of recovery for prenatal injuries, and delinking the fiction of the unborn child from the concept of the fetus. The stakes were many, among them the inscription of maternal–fetal division in tort law and a corresponding undermining of liberal governance of women.

Feminist scholars have argued that insurance was instituted to protect male heads of household against risk, with what might have been construed as women's domestic risks left uninsured (Richardson 2004). But the empirical site explored in the present work, the history of the legal fiction of the child *en ventre sa mère* and its attachments to risk reasoning, have had no single pattern of gender or social class effects, by turns serving and undermining women's gender interests. Women who gave birth after their husbands had died in occupational accidents gained from the child's entitlement to survivor benefits as the result of a new application of the child *en ventre sa mère* in occupational accidents legislation. Yet late twentieth-century civil litigation that attempted to assign each pregnant women a duty of care to her expected child also formed a new application of the child *en ventre sa mère* fiction and was universally condemned by feminists. At risk in the destabilizing of the child *en ventre sa mère* fiction was the birth threshold and the protections it provided women.

The child *en ventre sa mère* as legal fiction

The legal fiction of the "child *en ventre sa mère*" has existed for centuries in English common law and French civil law, the two legal systems that were received respectively into Anglo-Canadian and Quebec law. The fiction *en ventre sa mère* is currently entrenched in both Anglo-Canadian common law[6] and the Quebec *Civil Code*[7] with the effect of providing financial security for posthumously born children. In the event that the mother, or more typically the father, should die without a will or with a will containing no mention of future heirs from a pregnancy at the time of the testator's death, the posthumously born child will be read into the will as beneficiary. For this to occur, a number of conditions must obtain: (1) the child must be born alive; (2) the will must show no intent of exclusion; (3) the act of reading in must not be to the child's detriment. Estate law principles exist to guard the property of the posthumous child while s/he has not yet been born. A guardian (called a curator *in ventro* under Quebec civil law) may be appointed to care for the property that a child born after a parental death will inherit upon birth. Through child *en ventre sa mère* fiction, the unborn may own gifts and inherit property, with the ownership rights becoming effective at live birth (Weiler and Catton 1976: 643–645; Winfield 1942: 77).

Prior to the mid-nineteenth century, the fiction of the "unborn child" was used for the purpose of securing patrimony for a child not yet born. Although technically this fiction of estate law applied to both parties in a marriage, and presumably held for those cases where a woman died in childbirth but the child survived, in practice it was clearly intended for cases where a man died while his wife was pregnant. As married women's entitlements to legally own property were restricted in Western Europe and North America prior to the nineteenth century – although dower rights did exist and many qualifications to this bald statement might be made – it was primarily men who were in a position to draw up wills, and these were, as the phrase goes, "men of property" rather than peasants. It was only after the mid-nineteenth century, when posthumously born children gained entitlement to be included in survivors' benefits under fatal accident legislation, and subsequently under motor vehicle insurance, that the fiction of the child *en ventre sa mère* began to secure protection for non-propertied groups.

Under the common law, the "child *en ventre sa mère*" has been a concept of civil rather than criminal law. In common law regimes, civil and criminal law used separate ways of addressing the expected child, and for differing ends, the former concerned with patrimony and child support, the latter with homicide together with criminal assault and battery. As Cyril Means (1968: 422) has argued, within the classical common law the unborn were thought to exist any point after conception for the purposes of civil litigation.[8] Classical common law none the less applied a birth threshold to civil suits, since legal actions could be undertaken only where a child was born alive and viable. The criminal law, however, encompassed the unborn after quickening/animation; no criminal offences could be committed against the unborn prior to quickening. Since the introduction of legal commentaries with Bracton in the thirteenth century, the concerns of the criminal law have focused on drawing a line of demarcation beyond which the legal person can be said to exist for the purposes of homicide and abortion law, whereas those of the civil law were centred on inheritances and, from the nineteenth century, civil liability. The early legal commentators Bracton and Fleta maintained that killing the unborn after quickening constituted murder, but later commentators, Sir Edward Coke and Sir Matthew Hale, understood this act to be a misdemeanour rather than a felony and thus not a homicide (Gavigan 1984: 21). Early modern legal commentators held it difficult to know whether any given action had caused the death of the unborn after quickening, and the reported case law would appear to confirm this (Gavigan 1984: 22–23). Not surprisingly, prior to the statutory criminalization of abortion in 1803, reported English case law pertaining to abortion was, in Gavigan's (1984: 20) words, "very close to non-existent".

To say that the child *en ventre sa mère* is a legal fiction situates it within a figurative use of language recognized as integral to legal reasoning. Many fictions other than the child *en ventre sa mère* exist in contemporary Euroamerican legal regimes. They are familiar to lawyers, who become acquainted with them early in their training. Lon Fuller (1967: 53), author of the standard commentary on legal fictions in common law regimes, argues that their general use is "to escape the consequences of an existing, specific rule of law". They are conventionally understood as appearing in new and expanding areas of law, an initial attempt to extend previous law. Fictions are intended to "reconcile a specific legal result with some premise or postulate" (Fuller 1967:

51), and are rationalized as a common law device that permits the courts to create exceptions while not appearing to legislate.

Legal fictions are made with the knowledge of their rhetoricity, or as Fuller (1967: 9) put it, a fiction is a statement made "with a complete or partial consciousness of its falsity". At law, a fiction is understood as a departure from the "ordinary" or "natural" meaning of legal discourse, the taking on of a figurative sense.[9] The requirement that legal fictions be marked as tropic contains their rhetoricity as exceptional, enabling the denial of its presence elsewhere in legal discourse. Legal fictions thus mark a point at which legal discourse borders on discovering its own rhetoricity and reacts by trying to contain fictions within narrow limits or to eliminate them. Fictions throw into stark relief the performative/constitutive power of legal discourse.

Fuller's equation of fiction with "falsity" alerts us to a long history of unease with legal fictions dating to Gaius in ancient Roman law. Legal commentators have worried about the relation between legal fictions and truth, arguing that fictions are falsehoods and as such corrupt legal discourse (Birks 1986: 83–86; Olivier 1975: 88–90):

> Fictions of use to practice? Exactly as swindling is to trade.
>
> (Jeremy Bentham cited in Hamilton 1989: 1461)

> The use of fiction tends not only to impair, in a general way, reverence for truth; but also to diminish the respect which would otherwise be felt for the courts and for the law itself.
>
> (Smith 1917: 154)

Where *fictio* is formulated as separating law from truth, it is thought to undermine the functional relation of stunned reverence and the popular awe[10] thought necessary to the exercise of law. The commentaries on the relation between legal fiction and truth repeat the critique of rhetoric found in Ancient philosophy from the time of Plato's *Gorgias*: the opposition posed between rhetoric, understood as a practice of persuasion through manipulation/lying, and philosophy as a practice of truth-telling (Vickers 1988: 83–178). And fictions, even when they are not suspected of being outright lies, are persistently thought to have a bit of the sneak about them. Modern commentators, none more so than Jeremy Bentham (Hamilton 1989: 1459–1461), have objected to fictions as a covert form of law-making that properly belongs to the legislature.

Common law fictions have no unified normative intent, as can be seen from the following two examples. Example 1: In nineteenth-century England, the fiction of marital unity (that husband and wife are one at law) placed married women under what was called the "coverture" or protection of their husbands, severely restricting their property rights, and eliminating their legal capacity to enter into contracts or make wills (Doggett 1988). Example 2: One of the best known fictions of the twentieth-century common law, "attractive nuisance", was formulated in the context of a case that involved a child who was injured while playing on unfenced railway land. Technically the child had trespassed on the land and thus was not owed a duty of care by the railway company. None the less, by a fiction the child was held to have been invited on to the land, owed a duty of care as an invitee, and thus able to take action in court to recover damages (Fuller 1967: 66–68). The rhetorical force of fictions serves many ends.

Legal fictions are mutable. A fiction may be rejected by the courts or its meaning may be redefined, moving historically from a figurative to a non-figurative sense. Fuller (1967: 19) calls this latter process a movement from "live fiction" to "dead fiction", the latter lacking a sense of what he calls "pretense". Thus, for instance, the legal theorist Mary Condon (2001) has argued that what may be the most famous of legal fictions, the legal personality of the corporation, may now be losing its status as fiction. While particular fictions may be rejected or become dead metaphors, legal reasoning by means of fictions persists. Whereas modern forms of truthful expertise have been organized around the rejection of rhetoric and the alignment of truth with immanent, factual knowledge, the continuing presence of fictions within law points to an older form of expert reasoning that integrated rhetoric into its operation.

The legal fiction of interest here, the child *en ventre sa mère*, is conventionally dated to Roman law, with the following passages from the *Digest of Justinian* cited as its basis:

> 7. Paul, *Shares Which Are Allowed to the Children of Condemned Prisoners, sole book:* The fetus in the womb [*qui in utero est*] is deemed to be fully a human being, whenever the question concerns advantages accruing to him when born, even though before birth his existence is never assumed in favor of anyone else.[11]

> 26. Julian, *Digest, book* 69: For almost all purposes of civil law, children *in utero* are considered as existent beings. [*Qui in utero sunt in toto paene iure ciuili intelligunter in rerum natura esse.*] Even *hereditates legitimae* revert to them; and if a pregnant woman is taken prisoner by enemies, the child to be born has the right of *postliminium*, and accordingly follows the rank of his father or (as the case may be) his mother. Moreover, if a pregnant slave-woman is stolen, then although it is a purchaser in good faith who has possession of her when she gives birth, the child to be born is deemed stolen property and not subject to *usucapio*. By analogy to all this, it is also the case that a freedman, so long as a son of his patron might possibly be born, is subject to the regime which applies to those who have patrons.[12]

These articles from the *Digest of Justinian,* first published in 533 CE, mark the beginning point for standard legal citations related to the child *en ventre sa mère*, continuing into twentieth-century decisions. Case law quotes these articles from Justinian, frequently without citation, assuming it as common knowledge.[13] "*In rerum natura esse*", a phrase given the sense of completely born alive, comprises part of legal Latin to the present day. Although only the first sentence of Article 26 is customarily used, I have included the Article in full here because it shows relational questions that the fiction, *qui in utero est*, was meant to resolve: issues of inherited property, including property in persons (slavery), but also problems of inherited social status under conditions of warfare and paternal death. Modern case and statutory law dealing with the fiction of the "unborn child" has a primary concern with property, and a secondary one with inherited social status, *vide* the right of children born alive after a parental death to inherit peerage titles under English common law.

In the *Digest of Justinian* the *fictio legis, qui in utero est,* permitted limited legal interests to be bestowed prior to, and in anticipation of, a child's birth. This was not the only fiction found in ancient Roman law to facilitate the intergenerational circulation of property. Under the *lex Cornelia* a Roman citizen taken prisoner was no longer considered to be a citizen, and thus was considered to be without a will, for only Roman citizens could make wills (Fuller 1967: 61). By the *fictio legis Corneliae* the last moment of a testator's life occurred at the very moment of captivity, when the soldier was still a citizen. The fiction of the extra moment permitted the wills of captured soldiers to be recognized as valid. Concern for Ancient Roman property relations gave rise to temporal creativity.

The resort to fictionalizing the child *en ventre sa mère* may surprise us today, for in contemporary Euroamerican cultures it is common practice to speak of the "fetus". The child *en ventre sa mère* concept is not equivalent to our contemporary notion of the fetus, nor does the biomedical concept of the fetus have any place in legal reasoning about the child *en ventre sa mère* until the mid-twentieth century. As a discourse object, "fetus" has a history. The invention of perinatal mortality and prenatal risk assessment (discussed in the two previous chapters) serve as partial histories of how the fetus was formed culturally during the second half of the twentieth century. A medical concept dating to Greco-Roman medicine, "fetus" comprises the Latin translation of the Greek *embryon* (Dunstan 1990: 1–9; King 1990). However, the concept of the fetus found in our present is by no means equivalent to its Greco-Roman usage (Temkin 1956: xxxix), having undergone an epistemic transformation at the invention of modern biology. In biological thought humans did not so much procreate as reproduce, the latter characteristic shared across all forms of organic life, including plants; biology levelled the relation between humanity and other species (Jordanova 1995: 371–374). Some species such as the human gave birth to live offspring, and these took the form of a fetus prior to birth. The biological fetus, unlike the *non-dum* of the early modern period and the fetus of ancient medical thought, was a form of life considered to have its own immanent cause of organization that stimulated a sequence of standardized developmental stages *in utero*, with the fetus approaching closer and closer to viability, that is, the capacity for independent existence. Indeed it was only at the turn of the eighteenth century that visual representation of the unborn shifted from a symbolic to a realist register: from emblems (typically of boys) that were not understood as literally present in the womb, to anatomical drawings of the fetus (Duden 1999; 2002a: 69–82). The fiction, child *en ventre sa mère*, is thus part of an older system of symbolic representation of the unborn that preceded the invention of a realist, biological concept of the fetus at the turn of the nineteenth century.

In English common law, children born viable after a parental death have been able to inherit property since the 1587 decision in the *Earl of Bedford's Case*.[14] *Thelusson v. Woodford* (1805) represents the classic statement of the English common law about the inheritance of property and status titles under the child *en ventre sa mère* fiction. The decision in *Thelusson* confirmed that rights crystallize only upon live birth and viability.[15] The *Earl of Bedford's Case* and *Thelusson* are treated as part of Anglo-Canadian common law and cited in the decisions of the Supreme Court of Canada in many of the key constitutional cases during the 1980s and 1990s pertaining to the legal status of the fetus.[16] The applicability of the child *en ventre sa mère* fiction to Anglo-Canadian estate law was confirmed by the Manitoba Court of King's bench in *Re Charlton Estate* (1919)[17] and by the Supreme Court of British Columbia in *Re Sloan Estate* (1937).[18] These cases established that a child born alive and viable after his/her parent's death during pregnancy could by the fiction of the child *en ventre sa mère* be read into his/her parent's will, when not expressly mentioned there.

Anglo-Canadian common law does not extend to Quebec. The Quebec *Civil Code* and the common law do, however, share a common cultural understanding and similar provisions with respect to the child *en ventre sa mère* fiction. The *Civil Code of Lower Canada* (1866) remained in effect with amendments and additions until the official revision that came into effect in 1994.[19] Article 838 of the 1866 *Civil Code* permitted a child conceived but not born at the time of a testator's death to inherit:

> Persons benefited by a will need not be in existence at the time of such will, nor be absolutely described or identified therein. It is sufficient that at the time of the death of the testator they be in existence, or that they be then conceived and subsequently born viable, and be clearly known to be the persons intended by the testator.[20]

Several articles in the 1866 *Civil Code* conferred additional future property rights on a "child conceived but not yet born", with Articles 345 authorizing the appointment of curators to administer property on its behalf,[21] and Article 608 specifying that an inheritance can only devolve on those who are "civilly in existence", excluding "persons who are not yet conceived" and "infants who are not viable when born".[22] Article 771 made provision for gifts to an expected child, on condition that the "donee be conceived at the time of the gift or when it takes effect in his favor, provided he be afterwards born viable".[23]

In the nineteenth century, the fiction of the child *en ventre sa mère* found a new application in insurance law with the rise of insurance benefits under fatal accidents statutes and workers' compensation regimes. Winfield outlines the precedents set by the first UK statutes in this respect:

> Under the Fatal Accidents Act, 1846, it has been held that a child *en ventre sa mère* is considered as born for the purposes of sharing in the damages awarded for the death of the father caused by the wrongful conduct of the defendant; but until the child is born, its claim cannot be preferred. So too, under the Workmen's Compensation Acts, a child en ventre sa mère has been held to be a "dependant" of its deceased father in respect of compensation for his death.
>
> (Winfield 1942: 82–83)

The extension of the fiction from inheritance law to fatal accidents and other statutory insurance claims was associated with the workplace hazards of industrial capital (Doran 1996).[24] Through the inclusion of the child *en ventre sa mère* fiction in fatal accidents law, the legal fiction crossed a social class barrier, formally entering the lives of the working class.

Whereas case law pertaining to the child *en ventre sa mère* had for centuries been filled with disputes related to patrimonial land, leases, rents, gifts and titles, from the mid-nineteenth century cases brought under statutory benefits and insurance law told tales of workers dying in industrial accidents. *Giddings v. Canadian Northern Railway* (1920)[25] is the first reported Canadian case of fatal accident compensation being granted to a child born after a parent's death. The plaintiff was the widow of a railway worker crushed to death on the job by a railway car. She sought damages for herself and her child, born alive after the death of her husband. Both she and her child were declared "dependants" and granted survivor benefits under the *Act Respecting Compensation to the Families of Persons Killed in Accidents* (1909)[26] which was a fatal accidents-type legislation. The decision held that "[t]he child was unborn at the date of the death, but that appears to be immaterial so long as the action is for the benefit of the child".[27] In a later case, *Chapman v. C.N.R. and Parry Sound* (1943)[28] the Workmen's Compensation Board brought an action on behalf of the widow of a railway worker killed in a fatal collision on the job. The Board sued for damages on behalf of herself and her child, who was born after the accident. Damages were awarded to the woman under the Ontario *Workmen's Compensation Act*[29] and to the infant under the *Workmen's Compensation Act* in a subrogated claim under the *Fatal Accidents Act.*[30]

During the 1980s the scope of the child *en ventre sa mère* fiction under these statutory schemes for survivor benefits was extended in Canadian law to private motor vehicle insurance. In *Fitzsimonds v. Royal Insurance Co. of Canada* [31] (1984), a case was brought under the *Alberta Insurance Act* on behalf of a child whose father had died in a motor vehicle accident while his partner was pregnant.

The Alberta Court of Appeal confirmed the decisional ruling of the lower court that a "child *en ventre sa mère*" was by fiction considered a "dependent relative" within the meaning of the statute and, once born alive, able to claim death benefits.[32]

The legal fiction of the child *en ventre sa mère* has been linked to the provision of security through actuarial risk since the mid-nineteenth century in the UK and 1920 in Canada, authorizing financial support for the posthumously born children of workers dying in occupational accidents. The older functions of the child *en ventre sa mère* fiction in estate and child custody law[33] bore no relation to insurance, but its new applications in fatal accident legislation, workers' compensation and motor vehicle insurance were fuelled actuarially. Statute and case law recognized a child born alive and viable, as a "dependant" of an insured worker or driver of a motor vehicle and entitled to survivor benefits. The body of the deceased parent was insured as a capital, and the loss of that capital required compensation for those dependent on it. Survivorship interests and statutory accident benefits granted under the fiction of the child *en ventre sa mère* incited no public controversy. The new applications of the child *en ventre sa mère* led an obscure and recondite life within expertise, of no interest to anyone other than those lawyers, judges, insurance corporations and families involved in making claims for survivorship benefits.

The child *en ventre sa mère* and prenatal injuries

The applications of the child *en ventre sa mère* in fatal accidents legislation, workers' compensation regimes and motor vehicle insurance did not extend to the law of torts. Historically neither common law nor civil law torts allowed a child born alive to recover damages for wrongful conduct that had occurred before his/her birth. Internationally the first case that awarded damages for prenatal injuries was decided in 1933 under the Quebec *Civil Code.*[34] A new cause of action in the tort of negligence was recognized in this decision: a right of action to recover damages for prenatal injuries. It was not until 1972 that Anglo-Canadian common law, in tandem with a series of common law judgements and statutory changes in Australia, the UK and the United States,[35] permitted actions in prenatal injuries under the tort of negligence.

In claims for damages from prenatal injuries, a child who has been born alive sues a third party (i.e. someone other than his/her mother) for allegedly negligent acts during his/her mother's pregnancy that resulted in harm to the child's health after birth. The aim of the legal action is to provide the child's parents with funds for the additional costs of caring for a child with disabilities or illness. Suits based on prenatal injuries have to date been brought under the tort of negligence, although Fudge and Tucker (1993: 229) have argued that the intentional tort of battery might also provide a basis of liability for workplace prenatal injuries involving physical contact or toxins.

The site of injury under the fiction of the unborn child in fatal accident law, motor vehicle law and workers' compensation law is the *parent's* body. In claims for prenatal injuries the site of injury is the *child's* body: third parties are liable for negligent acts during pregnancy that become evident at birth. Both uses of the child *en ventre sa mère* fiction represent interventions to secure the welfare of children, but it is only in the case of prenatal injuries that compensation is for bodily harm done to *children.*

The law of torts has been constituted primarily as a vehicle for compensation and redistribution of losses, driven in the twentieth century by the availability of insurance. The tort of negligence has three components: a *duty* of care, a *breach* of conduct, and subsequent loss or *harm.* For legal purposes negligence is construed as a harm caused by a breach of conduct. One can only be negligent in tort to those owed a duty of care, which is "based on a relationship of proximity between parties requiring one person to take reasonable care for the protection of others" (Fudge and Tucker 1993: 231). The duty of care is subject to the standard of "reasonable foreseeability": how a person of "reasonable" care and foresight would act in similar situations.

The legal recognition of a right of recovery for prenatal injuries required that the cause of action be made consistent with the legal framework of negligence in tort law. The legal concepts of duty of care, reasonable foresight, and harm needed to be transposed to a type of case where breach (prior to birth) and harm (after birth) might be far apart in time and their causal connection difficult to show. Extending the fiction of the unborn child into tort law necessitated the following schema: people of *reasonable foresight* owed *a duty of care* to pregnant women, and *harm* to a child after birth could be linked to a *breach* (a negligent act) while the child was *in utero.* Finally, the cause of action necessitated that it in principle be possible to show that the *breach* was the *cause* of the *harm* appearing after birth.

Prior to the 1930s, no Canadian or other jurisdiction allowed actions for prenatal injuries. The key precedents, cited internationally, were US and Irish: *Dietrich v. Northhampton* (1884)[36] and *Walker v. Great Northern Railway Co. of Ireland* (1891).[37] Both cases denied compensation for prenatal injuries, although neither case was brought in tort law, with the claim in *Dietrich* brought under statutory law and *Walker* under the law of contract.

In *Dietrich,* a woman who was four to five months pregnant fell on a defective road in a Massachusetts town and miscarried; the baby died within minutes of birth.[38] The claim, brought against the town of Northhampton, had a basis in the statute law of Massachusetts, which permitted the next of kin of someone who had died as the result of a negligent act to sue the person who had acted negligently. On appeal to the Supreme Judicial Court of Massachusetts, the presiding judge, Oliver Wendell Holmes, reasoned that the case would, if allowed, set a precedent for permitting an infant, born alive, to maintain an action for prenatal injuries. As it was not certain that the child in *Dietrich* had been quickened, being "too little advanced in foetal life to survive its premature birth",[39] it was clear that the English criminal law standard, Coke's Rule, which applied to children after quickening, was not relevant to the case. This is a curious rendering of quickening as equivalent to foetal viability. Holmes went on to raise the question of whether "an infant dying before it was able to live separate from its mother" could be said to have become a person.[40] He held that a child not viable at birth could have no legal standing, and thus no legal claims could be brought on the child's behalf.[41] Justice Holmes reasoned that the "unborn child" formed part of the woman's body during pregnancy. In his concluding remarks in *Dietrich*, Holmes argued that "as the unborn child was a part of the mother at the time of the injury, any damage to it which was not too remote to be recovered for at all was recoverable by her".[42] As a result, if damage had occurred, it would be construed as pertaining to the woman, not the child.[43]

In the 1891 Irish case, *Walker v. Great Northern Railway Co. of Ireland,* a child ("A.W.") had been born with multiple permanent disabilities. During the pregnancy, her mother had been in an accident while in transit on the respondent's railway. Legal action was taken against the railway company, with the child, A.W., named as the plaintiff. The action was for personal injuries, a claim to recover damages for the alleged harm done to A.W.'s health by the accident in which her mother had been involved. A.W.'s lawyers argued that the company had a "duty to carry safely" and had been negligent in the performance of this duty.[44] A.W.'s statement of claim challenged the birth threshold: "A child *en ventre sa mère* is a person *in rerum natura,* and by the rules of the common and civil law is to all intents and purposes a child."[45] Counsel for the railway company argued that the plaintiff did not have the legal standing to launch the action as she had not been "*in rerum naturâ*" (completely born and viable) at the time of the accident, nor did the railway company have a duty of care to someone with whom it had no contract.[46]

The Irish Court of Queen's Bench decided that the railway company was not liable. The court reasoned that the company had made a contract with A.W.'s mother while she was pregnant, but had not entered into a contract with her "unborn child". Citing Justinian, *"Qui in utero sunt in toto paene jure civili intelligunter in rerum naturâ esse",* Chief Justice O'Brien speculated about the application of the civil law fiction to the law of contract:

> What has the carrier to say to this invisible person of the civil law? … The carrier saw the person he was going to carry. His duty was to that person. The carrier would be surprised to hear, while he was paid for one, that he was carrying two, or even three, for it might be a case of twins … He carries for hire. That is the fundamental account of his position and liability.[47]

As *qui in utero est* had not paid for the ride, there was no exchange and hence no contract. Nor could the railway carrier have had a duty of care to those who could not be seen, hence, "(t)here is no person and no duty".[48]

Justice O'Brien's (two justices by the name of O'Brien wrote judgements in *Walker*, Chief Justice O'Brien and Justice O'Brien) concurring judgement in *Walker* was sceptical that there could in principle ever be adequate evidence to substantiate proof of harm in prenatal injuries:

> there are instances in the law where rules of right are founded upon the inherent and inevitable difficulty or impossibility of proof. And it is easy to see on what a boundless sea of speculation in evidence this new idea would launch us. What a field would be opened to extravagance of testimony, already great enough – if Science could carry her lamp, not over certain in its light where people have their eyes, into the unseen laboratory of nature – could profess to reveal the causes and things that are hidden there – could trace a hare-lip to nervous shock, or a bunch of grapes on the face to the fright – could, in fact, make *lusus naturae* the same thing as *lusus scientiae*.[49]

The reasons for judgement revelled in contemplation of what Justice O'Brien regarded as, *prima facie*, preposterous. Evidence linking postnatal harm to prenatal breach would lead to "extravagance of testimony", hyperbole that could only be pernicious to the sobriety of contract law.

Justice O'Brien associated "extravagance of testimony" with injuries that were emotionally caused: "nervous shock" and "fright". In tandem with late nineteenth-century litigation over prenatal injuries, a number of fright-based actions connected with pregnancy were heard in courts of the British Empire and the United States (Chamallas and Kerber 1990: 82). Actions were launched in tort for harms suffered by women who had experienced miscarriages or stillbirths after being severely frightened, or who had witnessed physical injuries to their children. The references to "nervous shock" and "fright" in the 1891 *Walker* case echoed the decisions in the cases involving fright-based physical injuries, which showed a distrust of women as witnesses, hystericizing their testimony. The "extravagance" of women's testimony, particularly during or about pregnancy, was much feared in tort law (Chamallas and Kerber 1990), the costs of listening unbearable. In one sense then, both *Walker* and *Dietrich* are about the exclusion of women's risks from insurance coverage and a long history of disqualifying women's speech as dishonest. But in another sense the decisions in *Walker* and *Dietrich* affirmed the birth threshold in upholding the child *en ventre sa mère* as legal fiction.

The 1933 judgement of the Supreme Court of Canada in *Montreal Tramways*[50] broke with the established legal precedents set by *Dietrich* and *Walker*. The Court allowed a new application of the *fictio legis*, child *en ventre sa mère,* in the law of tort and granted a right of action in civil law torts for prenatal injuries. In its judgement, the Supreme Court of Canada accepted the risk arguments of medical expert witnesses as proof of fact that could link a negligent act during pregnancy to a child's injuries after birth. The case, which was brought under the *Civil Code of Lower Canada*, allowed a right of action for prenatal injuries before this was recognized as a cause of action in any common law jurisdiction. Internationally the landmark cases in common law regimes were *Bonbrest v. Kotz* (1946 – United States),[51] *Watt v. Rama*[52] (1971 – Australia) and *Duval v. Séguin*[53] (1972 – Canadian common law torts).[54]

In *Montreal Tramways*, the Supreme Court of Canada considered the case of a child born with club feet after her mother, then seven months pregnant, had fallen to the road while exiting a streetcar. The child's father launched the suit on the child's behalf. The action was brought under Article 1053 of the *Civil Code*: "Every person capable of discerning right from wrong is responsible for the damage caused by his fault to another, whether by positive act, imprudence, neglect or want of skill."[55] The Court interpreted the suit as involving two main legal principles. First, did the Articles in the *Civil Code* that permitted an "unborn child" patrimonial interests and guardianship by legal fiction once born alive also apply within the law of tort? Second, was there sufficient evidence to link postnatal injury with prenatal breach, that is, could causation be proved? With respect to the first question, the Court decided that Article 608, which permitted a child born alive and viable, if conceived at the time the succession devolved, to inherit property and accept gifts bestowed during gestation, and Article 345, which authorized the appointment of a curator to administer property for an unborn child, did extend to Article 1053. The "unborn child" was thus encompassed within the meaning of "another" in Article 1053. This interpretation enabled the extension of "unborn child" fiction from estate law to civil law torts. In the words of Justice Lamont writing for the majority, "the fiction of the civil law must be held to be of general application".[56]

The Supreme Court of Canada carried over into tort law the same legal framework as had previously governed the child *en ventre sa mère* fiction in property law:

> [A]lthough the child was not actually born at the time the company by its fault created the conditions which brought about the deformity to its feet, yet, under the civil law, it is deemed to be so if for its advantage. Therefore when it was subsequently born alive and viable it was clothed with all the rights of action which it would have had if actually in existence at the date of the accident. The wrongful act of the company produced its damage on the birth of the child and the right of action was then complete.[57]

These rules frame a set of conditions for a right of action to recover damages for prenatal injuries: the child must be "born alive" and "viable" in order to have legal personality and standing to sue. Harm is reckoned to begin at birth rather than at the time of the negligence, but breach is placed prior to birth. The awarding of damages is considered to provide for the additional costs of caring for an injured child. The person being sued is assumed to be a third party, that is, someone other than the woman who bore the child.

The decision in *Montreal Tramways* dealt extensively with the standard of proof in claims for prenatal injuries, relying on a variety of cases and commentaries in both the civil and the common law, though notably not citing *Donoghue v. Stevenson* (1932),[58] key to the standard of proof for causation in subsequently recognized common law torts. Nine medical expert witnesses testified at the trial, three for Léveillé and six for Montreal Tramways. The appellant, Montreal Tramways, argued that insufficient evidence existed to prove any causal connection between the accident in the seventh month of pregnancy and the child's subsequent disability. More than half the text of the majority decision examined the issue of causation and evidence, concluding that:

> under either the French or English jurisprudence the presumptions or inferences to be receivable as proof must be a deduction from established facts which produces a reasonable conviction in the mind that the allegation of which proof is required is probably true.[59]

The standard of proof adopted by the Court was one customary in Quebec civil law: "inference" based on "reasonable deduction" where cause and effect have not been directly observed.[60] The Quebec civil law standard of probable causation was used by Justice Lamont, speaking for the majority of the Court, to show that medical evidence linking postnatal injury to a prenatal accident was a reasonable inference, though not a certainty:

> In this case the cause which produced club feet cannot be demonstrated to a certainty and the law does not require it should be. It is simply a question of drawing an inference.[61]

He speculated that inferences were increasingly easy to make "with the advance in medical science".[62] The single dissenting voice in *Montreal Tramways*, Justice Smith, found the evidence insufficient since it did not permit inference by reasonable deduction; the evidence was a "mere guess".[63] Ultimately, the reasoning of the majority of the Court hinged on what it called "natural justice": to leave a child "bearing a very heavy burden of infirmity and inconvenience without any compensation therefor" was to accept "a wrong inflicted for which there is no remedy".[64] Ian Kerr (1997: 248), in an important article on the legal theory of prenatal injuries, has tellingly observed that, "[t]he court failed to provide a heoretical basis for its decision".

Montreal Tramways marked the first time that medical evidence about *fetal* development was accepted and summarized in a legal judgement that used the fiction of the child *en ventre sa mère* in its reasoning. Witnesses giving medical evidence for Léveillé and Montreal Tramways contested what kind of injury to a child's health might result from a fall in the seventh month of fetal development. Expert witnesses also debated what could be known about the effects of the accident on the particular patient. Dr Langevin, the main witness for the respondent, Léveillé, described the formation of fetal bones in the third trimester. His evidence was summarized by Justice Lamont in a passage I quote at length because it marks the first entry of medical evidence about the fetus into a judicial decision in tort law:

> Dr. Langevin, a gynaecologist and obstetrician, Professor at the University of Montreal, testified that in its mother's womb the child's members were in a flexed position and their malformation would be promoted by the absence of liquid in the uterine cavity which would cause the walls thereof to contract and flexing to increase. He further said that in the last months of pregnancy, particularly from the seventh to the ninth month, the calcification of a child's bones greatly increases; that during this period it requires twenty-two times more lime than during the first months, and that with the extra pressure caused by the contraction of the uterine cavity the chances of the bones calcifying in their flexed position become greater. He also said that when the pressure is found in the uterine cavity the probability is that a deformity will result. Dr. Langevin's conclusion was that while club feet may result from various causes, the only satisfactory explanation, in the circumstances of this case, was that the deformity resulted as a consequence of the mother's fall. In fact he said that scientifically there was no other explanation.[65]

This is the first passage in both civil and common law torts to treat the concept of fetal development as relevant to legal reasoning in tort. While prior legal decisions in tort had referenced a biomedical concept of the fetus (see *Dietrich* above), *Montreal Tramways* was the first decision to refer to and rely on medical evidence about the fetus and fetal development. Once medical evidence was accepted as proof of fact for damage claims involving prenatal injuries, the biomedical fetus in its developmental density was inserted into the judicial decision. In the passage summarizing Dr Langevin's evidence, biomedical reasoning appears in indirect discourse, his words being paraphrased and not directly quoted. Through indirect discourse medical knowledge found outside law was incorporated into judicial knowledge, transformed into legal fact. Transmitted into civil law torts via evidence as proof of fact, the fetus becomes proximate to the legal fiction of the child *en ventre sa mère*, a detailed biomedical representation of the unborn where the prior birth threshold had been silent.

Two differing ways of addressing the unborn, each from a separate historical period, were juxtaposed in *Montreal Tramways*: a realist, modernist, scientific fetus and an ancient *fictio legis,* the latter based on rhetorical address of the unborn. The resulting heteroglossia in the text of *Montreal Tramways* is breathtaking in its lack of historical awareness, poised at an innocent moment before fetus attempted to assimilate *fictio*.

The decision in *Montreal Tramways* turned and twisted on the acceptability of medical evidence as proof of fact. In one sense *Montreal Tramways* continued to reason legally in terms of a birth threshold for the legal subject; in order for suits to be launched on behalf of children allegedly injured prenatally, the child was required to have been born alive and viable. But where previous estate law, fatal accident and motor vehicle law had been concerned about parental deaths prior to a child's birth, and the care and status of children after birth, *Montreal Tramways* extended legal causation to the child before birth. The decision in *Montreal Tramways* reconfigured cause as beginning in a breach prior to birth and completed in harm crystallizing after birth. The birth threshold that is preserved in *Montreal Tramways* depended on medical evidence that reasoned in terms of a perinatal threshold. This was an unstable mixture at the time of judgement in 1933, one that became volatile in future decades with the consolidation of perinatal medicine.

Montreal Tramways remained an isolated, little cited, case until the 1970s. The first reference to *Montreal Tramways* in any Canadian jurisdiction, inclusive of Quebec courts, did not happen until 1956,[66] twenty-six years after it was originally decided. And while infrequently cited before the 1970s, since that time it has been referred to in common law decisions in Canada and elsewhere.[67] By the time of *Duval v. Séguin* (1972), the first case in Anglo-Canadian common law that allowed a right of recovery for prenatal injuries, the legal principles for the common law tort of negligence had fundamentally changed.

Prenatal injuries in Canadian common law torts

The facts of the case in *Duval* (1972) typify the perfectly awful fates of the children who are the subjects of liability claims for prenatal injuries. Three weeks after Huguette Duval had been injured in an automobile accident near Hawkesbury (Ontario) on a snowy winter day in 1969 when she was thirty-one weeks pregnant, she gave birth to a daughter, Ann. Her daughter had difficulty talking and walking, as well as physical and mental disabilities which Ann's neurologist attributed to brain injuries sustained during the accident. Ann's injuries left her cognitively impaired; it was believed that she was unlikely to finish secondary school. Justice Fraser of the Ontario High Court awarded Ann Duval damages for her prenatal injuries, the first time the property law fiction of the child *en ventre sa mère* was adopted into common law torts in Canadian jurisdiction. The recognition of the claim occurred, I will argue, at the price of further destabilizing the intelligibility of the child *en ventre sa mère* as a fiction.

By the time *Duval* was argued in the early 1970s, tort law had become a liability-expansive system of compensation. Where *Montreal Tramways* stood alone in allowing damages for prenatal injuries during the 1930s, the orientation of tort claims during the high tide of social welfare governance favoured awarding damages according to need in order to socially redistribute loss. Legal scholars (Conaghan and Mansell 1993: 97; Priest 1990) have argued that, during the period of growing welfare state/social governance, tort law broke with its previous focus on intentional fault. By the mid-twentieth century whether or not an individual intended to injure another was no longer key to determining liability and assessing damages. Judgements in the law of negligence moved to making corporations internalize the cost of harm and losses they had caused rather than having the harms and losses borne by the public purse or injured individuals (Priest 1990).[68] Corporations internalized cost through the payment of insurance premiums. Tort law was reconfigured as a technique for giving individuals compensation for harms and losses.

The recognition of prenatal injuries as a cause of action in *Duval* flows from these overall changes in the social function of tort law. The effect of the judgement in *Duval* was to require that the insurance industry accept a new type of claim against motor vehicle insurance. The courts assigned the cost of raising sick and disabled children to the plaintiff's insurer, with the insurer in turn redistributing its additional risk across its pool of subscribers. From the perspective of families, the compensation provided support for children with special needs. To insurers, recoveries for prenatal injuries expanded the risks for which their policy holders were potentially liable, with the body of the injured child considered actuarially as a capital loss needing compensation.

The new application of the child *en ventre sa mère* in common law torts was accomplished through two differing kinds of risk. First, the fiction of the child *en ventre sa mère* was carried into tort law on the slipstream of changes in twentieth-century negligence torts that transformed negligence into a mechanism of risk management for the redistribution of losses through compensation secured by insurance. This resulted in a broader judicial interpretation of those owed a duty of care. Second, the courts accepted expert biomedical evidence as proof of fact linking injury to perinatal risk factors. Judicial acceptance of medical evidence enabled a causal chain to be established between breach prior to birth and harm to a child after birth.

Duval generalized the child *en ventre sa mère* from property law to common law torts, but, as Ian Kerr (1997: 249) has emphasized, the decision "awarded damages for pre-natal injuries without *expressly* employing a legal fiction" (Kerr's emphasis). The trial judge, Justice Fraser, ruled that the cause of action for prenatal injuries is complete when a child is born alive with injuries arising from an accident during pregnancy: a transposition of the live-birth rule from property law to common law torts. In order to recognize a cause of action, Justice Fraser needed to make prenatal injuries conform to the legal reasoning found in the law of negligence: a negligent act must be incurred by someone in a class of people owed a duty of care, and that breach must be the cause of loss or harm to that person. To do so his judgement relied extensively on the landmark decision in the House of Lords, *Donoghue v. Stevenson* (1932).[69]

Practising lawyers often recall being taught that the tort of negligence began with *Donoghue*. However, scholars as diverse as Winfield (1926) and Horwitz (1977), though differing markedly in their analyses and reasoning about historical context, date the tort of negligence to the nineteenth century. *Donoghue* transformed the legal framework of negligence torts, making corporations liable for products that harmed public safety, but the tort of negligence existed long prior to this decision. In *Donoghue*, a beverage manufacturer was held liable for injuries suffered by a woman who became sick after drinking a beverage produced by the company; the drink contained decomposed snails, the most famous snails in legal history. The manufacturer did not intend to cause the plaintiff harm, not knowing of her particular existence at the time of manufacture, but did intend to sell its beverages to consumers. The judgement in *Donoghue* reasoned that consumers of the beverage had been a reasonably foreseeable class of people for the manufacturer. The standard of reasonable foreseeability articulated in the decision limited the duty of care to those classes of people that someone can "reasonably" anticipate being affected by the act (Conaghan and Mansell 1993: 38–39 and 49–52). *Donoghue* held that people and corporations can be held accountable in negligence for actions that they ought reasonably to have known would cause loss or harm even if they did not intend to cause it.

The "rule of proximity" in negligence torts configures the allowable distance in space and time between breach and harm. The legal subject cannot be held liable in negligence for acts where the space and time between initial breach and subsequent harm are reckoned "remote"; this means that the consequences of an act cannot be practically anticipated by the legal subject. Beginning with *Donoghue* and continuing in succeeding decisions,[70] the rule of proximity was relaxed to permit greater distance in space and time between the probable cause of an injury and the resultant harm. In *Donoghue*, the breach (placing the snails in a beverage bottle) and harm (sickness probably due to drinking a beverage containing essence of snail) were distant in time and place. When the rule of proximity was extended to cover greater time–space distance, the responsible legal subject was expected to be able to reasonably foresee the consequences of her/his actions over much longer periods of time and greater spatial distances.

It has become conventional to interpret the decision in *Donoghue* as having multiplied the classes of persons to whom a duty of care is owed by extending the temporal and spatial reach of proximity. *Donoghue* proliferated the neighbours surrounding the legal subject. The temporal and spatial extension to the rule of proximity expanded the compass of reasonable foreseeability, the normative expectation that legal subjects anticipate the practical results of their actions. The increased temporal and spatial depth given to reasonable foreseeability led to an expansion of the classes of those owed a duty of care.

The *Donoghue* decision altered the governance of uncertainty in tort law. As legal subjects cannot typically numerically calculate the outcomes of their actions, *Donoghue* did not require that subjects govern their conduct through risk, but through incalculable uncertainty. In the changed framework of negligence after *Donoghue*, the responsible legal subject was conceptualized as an actor envisioning the consequences of his/her actions for multiple classes of persons in future times and spaces. This marked a change in the normative construction of the legal subject, but also a reconfiguration of uncertainty to include increased temporal and spatial range: a far-reaching future populated by more classes of subjects owed duties.

The extension of reasonable foreseeability and the rule of proximity developed in *Donoghue* formed the preconditions for the recognition of prenatal injuries in *Duval.* Harm after birth was no longer "too remote" to have been anticipated by the legal subject at the time of the breach during pregnancy. Injury to a child born alive could be configured as caused by a wrongful act during pregnancy. But exactly which class of persons was owed a duty of care in actions for prenatal injuries was culturally unclear. The child after birth? The unborn/fetus before or during birth? Could a child *en ventre sa mère* fiction be harmed? In the Canadian jurisdiction the biomedical fetus began its occupation of the child *en ventre sa mère* fiction with *Duval.*

In *Duval* Justice Fraser cited *Donoghue* to establish both that the child *en ventre sa mère* fell within the scope of reasonable foreseeability and that cases of prenatal injuries were covered by the rule of proximity. The decision in *Duval* quoted the famous passage from Lord Justice Atkin in *Donoghue* (1932) that had introduced what later came to be called the "neighbour principle" as the standard of reasonable foreseeability:

> The rule that you are to love your neighbour becomes in law: You must not injure your neighbour; and the lawyer's question, Who is my neighbour? receives a restricted reply. You must take reasonable care to avoid acts or omissions which you can reasonably foresee would be likely to injure your neighbour. Who then, in law, is my neighbour? The answer seems to be, persons who are so closely and directly affected by my act that I ought reasonably to have them in contemplation as being so affected when I am directing my mind to the acts or omissions which are called in question.[71]

Duval used the "neighbour principle" to argue that drivers should anticipate that there will be pregnant women in cars on the road, and therefore drivers owe a duty of care to children *en ventre sa mère:*

> Ann's mother was plainly one of a class within the area of foreseeable risk and one to whom the defendants therefore owed a duty. Was Ann any the less so? I think not. Procreation is normal and necessary for the preservation of the race. If a driver drives on a highway without due care for other users it is foreseeable that some of the other users of the highway will be pregnant women and that a child *en ventre sa mère* will be injured. Such a child therefore falls well within the area of potential danger which the driver is required to foresee and take reasonable care to avoid.[72]

The reasoning in *Duval* sharply contrasts with the 1892 decision in *Walker*, which had mocked the suggestion that a railway company could reasonably be expected to imagine pregnant women among its customers or that evidence could be brought to link prenatal negligence with postnatal harm. "Ann's mother", Justice Fraser maintained, was in a class of people owed a duty of care and subject to "foreseeable risk", and so was Ann.

Justice Fraser cited *Donoghue* again when considering the question of proximity:

> Since *M'Alister (or Donoghue) v. Stevenson*, *supra*, and the many later cases which have followed it, the law has been clear that it is unnecessary that the damages coincide in time or place with the wrongful act or default.[73]

The *Duval* judgement applied the extended reach of proximity to connect a breach during pregnancy with post-birth harm. The right of a child to sue for prenatal injuries remained subject to the live-birth rule. For the purposes of tort law, the cause of an action was not considered "complete" until after a child has been born.[74] As Fudge and Tucker note: "In effect, the court creates the legal fiction that the defendant's negligence while the plaintiff was *in utero* does not cause damage until the child's birth" (Fudge and Tucker 1993: 252).

The live-birth rule carries the implication that the unborn has no legal standing and cannot initiate an action in tort. This was confirmed in *Dehler v. Ottawa Civic Hospital* (1979)[75], *Seede v. Camco Inc.* (1985),[76] followed by the Supreme Court of Canada in *Winnipeg Child and Family Services (Northwest Area) v. G.(D.F.)* (1997)[77] (discussed in the next chapter). In *Seede v. Camco*, a woman who had suffered a miscarriage after a car accident sued on behalf of herself and other family members for the loss of care and companionship due to the accident. The Supreme Court of Ontario dismissed the action, stating that:

> In Ontario there is no right of action in an unborn child to recover damages nor in favour of an unborn child unless the child is born alive. On the other hand, when the unborn child becomes a living child through birth and suffers damages as a result of prenatal injuries caused by the fault or the negligence of another the cause of action is completed.[78]

Neither the Canadian common law nor the Quebec civil law recognizes wrongful death actions for fetuses or stillbirths.[79] Miscarriages and stillbirths occurring as the result of a negligent or intentional action do not provide a basis for action in damages.

In *Duval* Justice Fraser construed the damage as occurring after birth, but "it was the *foetus* or *child en ventre sa mère* who was injured",[80] a curious equivalence given that "fetus" represents a biological entity and "child *en ventre sa mère*" a fiction. As in *Montreal Tramways,* recognition of a right of action for prenatal injuries was conditional on finding some way of supplying evidence that proved fact. Any claim in negligence specifically requires proof of causation linking the defendant's actions with harm or loss for the plaintiff. We saw above how courts in the late nineteenth century thought such proof impossible, even ridiculous to contemplate, in actions based on prenatal injuries. Proof of fact in the claims arising from prenatal injuries is "almost totally dependent on expert testimony to establish causation for the obvious reason that any external manifestation of injury to the fetus cannot be discerned until it is delivered sometime later" (Lintgen 1962: 565). Suits based on prenatal injuries require the evidence of expert witnesses who are health care providers, chiefly physicians, to give an opinion as to whether an event during pregnancy resulted in ill-health to a child after birth. This task requires physicians and other health care personnel to link the bodily substance of the fetus with that of the child across the perinatal interval. From the mid-twentieth century, expert witnesses conceptualized the link between the fetus prior to or during birth and the child after birth in terms of probabilities and risk judgements.

By the time *Duval* was heard in 1972, a series of decisions internationally had followed *Montreal Tramways* in admitting medical evidence in claims arising from prenatal injuries.[81] Justice Fraser in *Duval* allowed medical evidence as proof of fact to show cause.

> Some of the older cases suggest that there should be no recovery by a person who has suffered prenatal injuries because of the difficulties of proof and of the opening it gives for perjury and speculation. Since those cases were decided there have been many scientific advances and it would seem that chances of establishing whether or not there are causal relationships between the act alleged to be negligent and the damage alleged to have been suffered as a consequence are better now than formerly.[82]

Two physicians gave evidence on behalf of Ann Duval, with differing opinions. The general practitioner who had attended the birth thought Ann's condition occurred independently of the car accident,[83] but deferred to the neurologist who linked Ann Duval's mental and physical disabilities with injuries caused by the car accident when her mother was pregnant.[84] The neurologist was uncertain as to what caused the partial fusion of bones in Ann Duval's skull, but believed she suffered from brain damage that was caused by the accident independently of the fused bones.[85] At trial the neurologist testified:

> My final opinion was that this baby had suffered from a head injury during the time she was being carried by her mother at the accident of February the 16th, and that this accounted for the fact that the mother felt little movement for some period of time after the accident, and it also accounted for the fact that the child had damage to motor cells in the brain, resulting in the leg problem that we have described, and also resulting in … some degree of retardation … I do not know whether the accident had anything to do with the skull abnormality, but I feel that in all probability it did have to do with the damage to the motor cells in the brain, and also had to do with her mild to moderate retardation.[86]

The medical experts reasoned in terms of the probability/risk of the child's injuries having been caused by the automobile accident, which the Court interpreted through the legal standard of probable cause. In the medical testimony the association between child and the unborn was conceptualized as one of biological continuity. The association was established through probable pathophysiological sequences about brain tissues and cranial bone formation rather than a temporal fiction linking the born with the unborn.

Duval set a precedent in applying the legal fiction, child *en ventre sa mère*, to Canadian common law torts. Justice Fraser was careful to note that he recognized no claim for or against fetal personhood. He reasoned that, although an injury may be sustained prenatally, a claim can only be advanced after a child has been born alive (the live-birth rule):

> In the instant case the plaintiff sues, as a living person, for damages suffered by her since her birth as a result of prenatal injury caused by the fault of the defendant. In my opinion she is entitled to recover such damages. I refrain from expressing any opinion as to what, if any, are the legal rights of *a child en ventre sa mère or of a foetus*. Many difficult problems in this area of the law remain to be resolved.[87] (emphasis added)

These comments are striking in historical terms for their equation of the child *en ventre sa mère* with the biomedical fetus. "Foetus or child *en ventre sa mère*" have become part of the same paradigmatic set, each term capable of substituting for the other:

> While it was *the foetus or* child *en ventre sa mère* who was injured, the damages sued for are the damages suffered by the plaintiff Ann since birth and which she will continue to suffer as a result of that injury.[88] (emphasis added)

Duval addresses the child *en ventre sa mère* as capable of sustaining injury:

> If a driver drives on a highway without due care for other users it is foreseeable that some of the other users of the highway will be pregnant women and that a child *en ventre sa mère* may be injured.[89]

Motorists are expected to know the fetus is on the road, one of the foreseeable class of people carried in automobiles. But how might a fiction be injured? One could no more injure the legal fiction of the child *en ventre sa mère* than one could run over Mickey Mouse. The rhetorical address at law of the child *en ventre sa mère* has become unsettled.

The *Duval* judgement shifted the meaning of the child *en ventre sa mère*, which appears in the decision as a nominal group that acts as a receiver (the acted upon) in verbal processes of doing and acting (material processes)[90]; the *fictio* of the child *en ventre sa mère* is thus configured as capable of being injured. The *Duval* decision surrounds the child *en ventre sa mère* with the same legal rules that had for centuries framed this concept in both estate and accident law, but it has changed the meaning of the concept, shifting it into a realist register from a fictive one. The concept of the child *en ventre sa mère* in *Duval* falls midway between fiction and fetus.

The movement of the child *en ventre sa mère* fiction from property law to the law of negligence strained at its intelligibility as a fiction. My argument here parallels one made by the legal scholar Ian Kerr (1997) who, in an influential article cited by the Supreme Court of Canada,[91] shows that the fiction of the child *en ventre sa mère* was extended beyond its theoretical limits as a legal concept in *Duval*. Kerr (1997: 252) argues that, in *Donoghue* "the notion of a foreseeable plaintiff is *restricted to persons*" (Kerr's emphasis). Because the child *en ventre sa mère* in *Duval* was not yet born at the time of the accident, then, for the purposes of common law, the child *en ventre sa mère* "was not yet a *person* at the time of the collision" (Kerr 1997: 253). Kerr thus maintains that Justice Fraser's reasoning in *Duval* was in legal error: a child *en ventre sa mère*, being not a legal person, was not covered by the neighbour principle.

The category error in *Duval* – the confusion of the unborn child as a *fictio legis* with the biomedical concept of the fetus some fifteen centuries after Justinian – is an instance of a process that Marilyn Strathern (1992: 137–139) has called "literalisation": a drive to make the taken-for-granted explicit that is characteristic of our present. Literalization refers to the restless search for the "really real", a scientifically rendered "real" held to be prior to and the foundation for social and cultural classification (Strathern 1992: 137). And thus a legal fiction became a biological fetus.[92]

Legal fiction, risk and reality effects

The legal fiction, child *en ventre sa mère*, has been used to implant the inheritance of property, succession to social status, and the support of children. Prior to the nineteenth century, the child *en ventre sa mère* was situated in estate law, reading children born after a parental death into wills. But in the nineteenth and twentieth centuries, the legal framework of the child *en ventre sa mère* was for the first time conjoined with governance, bonding with the risk techniques of insurance. Its new applications had insurance as their condition of existence. Case law and statutes fixed the parameters of insurance compensation to include claims made on behalf of posthumously born children and children with injuries from prenatal accidents.

Situated at the seam of law and governance, the recognition of a right of recovery for prenatal injuries acted to assist health at the threshold of the living subject. Third parties, typically motor vehicle drivers, who injured women during pregnancy were made liable for injuries to children who were born alive. Prenatal injuries in the law of negligence had a deterrence logic of conserving the health of the living subject prenatally, although of course the intent of anyone filing a suit has been to obtain money in order to care for sick/disabled children. Medical knowledge made the bodily effects of prenatal injuries knowable at law, putting into discourse the impact of events prior to birth on health after birth. The new application of the child *en ventre sa mère* fiction was inherently dependent on medical knowledge of the perinatal threshold.

After *Donoghue* the responsible legal subject in the tort of negligence was required to govern her/himself prudently under conditions of uncertainty, that is, under conditions where the outcome of actions cannot be calculated, where probabilities are unknown.[93] In negligence torts, this configuration of the legal subject depended on the antecedent existence of insurance; the duty of care in relation to prenatal injuries gave rise to liabilities that were compensated by motor vehicle insurance. Governance of the legal subject through uncertainty is thus dependent on insurance as a technology of risk.

Risk techniques also entered legal actions for damages related to prenatal injuries at the level of proof of fact. Proof of fact in suits for prenatal injuries systemically requires expert evidence calculating the probability that a negligent act which occurred during pregnancy caused a child's sickness/disability after birth: a judgement of clinical risk. But, as I argued in Chapter 2, clinical risk judgements fall in technical uncertainty since risk to an individual cannot be calculated. Risk discourse has been used in litigation, across diverse branches of law to reconfigure the conception of the responsible legal subject as self-governing in and through uncertainty. In litigation discussed in the present chapter the legal subject was addressed as having the duty to foresee the *risks* of a particular action for others rather than simply having a practical sense of what it is to act under conditions of uncertainty. The movement of risk and uncertainty thus has symmetry, with medical evidence of risk being formally uncertain and the governance of the responsible legal subject being by turns modelled on both uncertainty and risk. In this process the relation between risk and uncertainty became analytically unstable.

The decisions in *Montreal Tramways* and *Duval* excited no public controversy. Neither case had any interveners, unlike the later, highly politicized cases in criminal, constitutional and tort law during the 1980s and 1990s that involved claims to the legal personhood of the fetus. Until the principle of maternal immunity in tort law was attacked during the latter half of the 1990s, no one (other than the insurance companies named in the various precedent-setting suits) objected to a child who had been born with health problems receiving compensation that would assist the child's family with the additional costs of care. The principle of maternal immunity in actions for prenatal injuries prevented women from being sued for what later came to be called "prenatal negligence", either on behalf of the fetus they were carrying or by their children for antecedent behaviour during pregnancy.

For centuries there had been no question of maternal–fetal conflict arising through the child *en ventre sa mère* because no fetus existed in this form of legal address. As mentioned above, the concept of the fetus has been found in academic medical discourse since Ancient Greece and Rome, but it was not construed biologically as a form of life with immanent principles of organization. After the formation of modern biology during the first years of the nineteenth century, the biological fetus was conceptualized as having developmental stages, becoming progressively more organically complete and fit for living outside the mother as gestation advanced. Unlike the legal fiction of the child *en ventre sa mère,* the fetus in the nineteenth and twentieth centuries comprised part of a realistic system of representation: "fetus" represented an entity regarded as anterior to discourse. The fetus in scientific realism is considered to be a being that exists independently of the human mind, but scientific perception is none the less thought able to grasp this anterior reality, which is regarded as having stable essences. The biological fetus was never a fiction, although it was a symbolic construct.

In the twentieth century the fiction of child *en ventre sa mère* was destabilized when the fetus entered tort law via medical evidence. In *Montreal Tramways*, the legal fiction of the child *en ventre sa mère* was for the first time placed in conjunction with a medically mediated knowledge of the fetus. Attaching a realistic concept of the fetus as proof in evidence to the legal fiction of the child *en ventre sa mère* was not necessarily incoherent. However, the proximity of the child *en ventre sa mère* to the fetus did contain the danger of a realistic reductionism that would equate the two concepts and collapse the birth threshold of the legal subject into the perinatal threshold of medicine. Risk judgements are probabilistic, but they are based on epistemological realism in representation. The fetus in medical opinions about risk during pregnancy is not put into discourse as a fiction or symbol, but as a referential concept: a body with material processes knowable through scientific knowledge.

The recognition of a right of action to recover damages for prenatal injuries created confusion: the possibility of displacing the fiction of the child *en ventre sa mère* and substituting the fetus, paralleling proposals made before the courts in criminal and constitutional law from the late 1970s in Canada. During the last years of the twentieth century, child welfare agencies identified the child *en ventre sa mère* with the fetus in litigation brought under tort law. The effect of such a reconstruction in legal reasoning would have been to require that pregnant women govern themselves to avoid foreseeable risk to the fetus during pregnancy so as to avoid suits in negligence. The problematization of the child *en ventre sa mère* as a fetus proposed restrictions on women's conduct during pregnancy for the sake of fetal security. This was an argument for a fundamental revision of tort law, one that challenged the boundary between law and medicine. It was a proposal of such magnitude that it was rejected by Canadian courts on appeal through an affirmation of the birth threshold. Biomedical evidence had introduced reality effects into the child *en ventre sa mère/qui in utero est* that made reasoning through fictions enormously difficult. Like Alice in Wonderland (see the second epigraph to this chapter), that sturdy English empiricist, belief in impossible things had become strained by lack of practice as a profound cultural distaste for the symbolic settled over the human body, conceived as it were biologically.

5 Child welfare at the perinatal threshold

Making orders protecting fetuses

What is essential for the purposes of thc case at bar is this: Aboriginal communities, their leaders and their health organizations are actively engaged in responses to the fetal health problem, but resort to involuntary confinement and treatment has not been endorsed by Aboriginal peoples. In every instance, community health approaches and innovative voluntary treatment strategies are being formulated and implemented, although clearly more support from Canadian governments is needed.

(*Winnipeg Child and Family Services (Northwest Area) v. G.(D.F.)*, [1997] 3 S.C.R. 925)
(*Factum of the Intervenor* Women's Health Clinic Inc., Métis Women of Manitoba Inc., Native Women's Transition Centre Inc., Manitoba Association of Rights and Liberties to the Supreme Court of Canada at para. 46)

I wish there was some way that you could help the unborn without forcing women into treatment ... Forcing women into treatment is far different than providing them with treatment.

(G. quoted in Rutman *et al.* 2000: 68)

The perinatal threshold was constituted in medicine as a way of reducing mortality in the days after birth. When the duration of the perinatal as one continuous biological interval before, during and after birth was introduced into law, the prior legal address of pregnancy and the beginning point of the legal person were put into question. The problematization of the threshold of the legal subject crossed many branches of law: constitutional, criminal and civil, statutory and case. The basic question at issue was the deceptively simple one of demarcating the beginning point of the legal subject, specifically whether the demarcation occurred before, during or after birth. The parties posing this question and the surfaces from which it emerged had quite different motivations, ranging from the Christian Right who opposed all forms of abortion on ethical grounds, to child welfare authorities who wished to secure the health of children in their charge. The attempt to give the perinatal threshold legal force constituted an event in biopolitics: a competition between the birth and perinatal thresholds. Inside and outside the courts the threshold of the living subject was debated in terms of "the fetus", the particular discourse object to which the perinatal attached before and during birth. Fetal personhood was an effect of the perinatal threshold received into law. Whereas the legal fiction of the child *en ventre sa mère* adhered to the birth threshold, vesting rights in a child only after s/he was born alive, at law the perinatal threshold resulted in proposals to constitute the fetus as a legal person with rights prior to birth. The interpretation of the fetus as a legal person would have created the now familiar but historically unprecedented legal situation of two legal persons sharing the same body (fetus and pregnant woman), with potentially conflicting rights.

The present chapter examines a line of cases brought during the past two decades of the twentieth century by Canadian child welfare authorities seeking to enlarge their jurisdiction to include the fetus. The mandate of child welfare agencies in child welfare law had been defined by the birth threshold: the legal status of "child" was conferred at birth. In order to accomplish an extension of their jurisdiction to the fetus, child protection services sought to replace the birth threshold with a perinatal one that established continuity between the health of the child and that of the fetus. A number of legal actions in pursuit of these goals were brought under child welfare law, mental health law, *parens patriae,* and tort law. Provincial child welfare agencies launched these cases with the aim of securing the future health of children in their custody by preventing risks to fetal health during pregnancy. They had no quarrel with women's access to abortion. Their concerns lay entirely in finding some and any legal means of securing children's health after birth by acting on women during pregnancy and labour.

The surfaces of emergence related to the problematization of the legal address of pregnancy were politically and socially disparate. *Borowski*[1] and *Daigle*[2] involved anti-choice challenges to abortion access on the grounds that the fetus was a human person. The plaintiffs in *Borowski* and *Daigle* submitted extensive biomedical evidence with the intent of proving unique human life from conception. The biomedical evidence was argued in support of the argument that it was "morally wrong to deny the legal status of personhood to anyone who is biologically human" (Stouffer 1994: 62). In these cases biomedical and scientific evidence was employed to make claims about the moral/legal status of the fetus, not about fetal health risks. Where instead concern falls on the amelioration of health, the fetus is given value by reference to the health of the child, an argument quite separate from making existential claims about the moral status of the fetus *per se.*[3]

Since the time of Peller and Baird, perinatal mortality has been linked with poverty; differences in comparative perinatal mortality rates within nation-states empirically coincide with social divisions based on wealth. Those populations characterized by high comparative perinatal mortality rates become identified by health governance as driving up national and regional perinatal mortality rates, and accordingly are made the targets of remedial efforts to lower their perinatal mortality rates. Health governance analyses populations with higher comparative perinatal mortality rates as pools of perinatal risk, developing policies to intervene in the control of their perinatal risk factors.

In the Canadian context, a series of federal government studies during the 1970s identified Aboriginal peoples living on reserves in the Canadian North as having significantly higher perinatal mortality rates than the pan-Canadian one. Given the significance attributed to the perinatal mortality rate as an indicator of national health, the perinatal mortality rate of Inuit and Aboriginal women could not be governmentally tolerated. The governmental response was to authorize a policy of universal hospital confinement for northern Aboriginal women during childbirth so as to reduce their exposure to perinatal risk. During the 1980s Inuit and Aboriginal women were evacuated from the Canadian North to hospitals sometimes far from their homes, separated from their families later in pregnancy, during and after birth (O'Neil and Kaufert 1990; Kaufert and O'Neil 1993). In this manner Aboriginal women enrolled in the history of the perinatal threshold, airlifted to lonely births and high obstetrical intervention rates in Southern Canadian hospitals.

Efforts to reduce the perinatal mortality rate among Aboriginal peoples on reserves in Northern Canada coincided with the invention of "fetal alcohol syndrome" and "fetal alcohol effects" during the 1970s, after which alcohol and a series of other substances such as crack cocaine and glue came to be identified as perinatal risk factors. The proliferation of perinatal risk factors drew the attention of child welfare authorities, whose mandate was in part to protect the health of children, and whose empirical field of intervention involved marginalized populations, particularly the children of women who were unemployed, single, judged mentally ill or using substances considered addictive. The growing importance placed on perinatal risk factors led child welfare authorities to conceptualize their mandate to include preventive action during pregnancy to protect child health among Aboriginal and underclass women, the chief subjects of child protection services. The subjects of child protection services came to be perceived through a screen of perinatal risk factors.

In the litigation considered here, the relation between risk and uncertainty was rendered unstable, in some instances debated as such among the parties to the litigation and in the legal judgements. Medical evidence of fetal health risks falls within the criteria for biomedical certainty when given in epidemiological form as the report of the relation between exposures and outcomes over a study population or as the metanalysis of multiple studies. As I showed in Chapter 1, the outcome from an exposure to an individual cannot be known in advance, that is, the results of exposures to individuals are prospectively uncertain. This can be illustrated by reference to the literature on alcohol use during pregnancy. It is accepted in the biomedical literature that not all children who have alcohol exposure *in utero* will have fetal alcohol syndrome/effects (FAS/FAE). In cases of women who regularly drink what studies operationalize as excessive amounts of alcohol during pregnancy, roughly 5 per cent of children will have fetal alcohol syndrome (Armstrong 2003: 7). Clinical risk judgements, whether made in a physician's office or in a court of law, are technically uncertain; nineteen of twenty women who drink a lot during pregnancy will have babies without FAS/FAE and doctors are not able to tell which children will be affected and which will not. Health care professionals know this, and they further consider fetal alcohol effects/syndrome a difficult diagnosis to make clinically. One of the problems at the clinical level is that doctors are predisposed to see FAS/FAE among the socially excluded (Armstrong 2003: 5), leading to the confounding of poverty and ethnicity with diagnosis. From the perspective of the history of the present, my point is not to discredit risk estimates about fetal health on evidentiary grounds, but rather to note that within health governance such estimates are contested and regarded as uncertain.

Sheila Jasanoff, a leading scholar of scientific evidence at law, has shown that scientific proof and legal proof, scientific fact and legal fact, are normatively and procedurally distinct. When expert evidence is given in litigation, scientific knowledge is reconstructed for legal purposes (Jasanoff 1995: 42). Jasanoff has shown how in American legal decisions in tort dealing with ecological disasters, judicial fact-finding has consistently preferred clinical evidence over epidemiological studies, privileging evidence of the sick body over population-based quantitative evidence of risk (Jasanoff 1995: 123). However, at issue in the child welfare litigation discussed herein was not the judicial ranking of biomedical knowledge, but whether any kind of biomedical knowledge was judicially relevant as the legal implications of admitting such evidence would have taken apart the distinction between the born and the unborn, potentially across all branches of law.

The secondary literature on risk has for the most part interpreted it as intrinsically liberal and democratic, a technology for distributing losses across the members of a risk pool that citizens freely accept, either through private insurance or through the democratic practices of states in social insurance schemes (Ericson, Doyle and Barry 2003; Ewald 1991 and 1986; O'Malley 1996). More recent studies have noted incitements to voluntarily engage in risk – to "embrace risk" and uncertainty through stock market investments or extreme sports (Baker and Simon 2002; O'Malley 2000 and 2004). Others, particularly those investigating the tender mercies of risk in the criminal justice regime through the classification of prisoners or parolees (Hannah-Moffat 1995: 149–155; Kemshall 2002) have noted practices of involuntary risk assessment.

Authoritarianism has been consistently present in the modern project of conserving the health of populations. It takes many forms, one being intervention in the health of subpopulations thought to endanger the health of the many. The central analysis of vital authoritarianism in Foucault's writings occurred in the last section of the *History of Sexuality*, Volume 1, "Right of Death and Power over Life" (Foucault 1978: 133–159), where he discussed the genocide of the European Jews under German fascism as an instance of the permanent capacity in modern biopolitics to kill subpopulations, particularly racialized groups, lethally fantasized as representing a threat to popular health.

The form of authoritarianism deployed in the child welfare cases analysed here operated within law and through risk; evidence about fetal risk attempted to align law with security. In attempting to institute a perinatal threshold at law, child protection agencies were motivated by a concern to reduce the numbers of ill and disabled children permanently in care, children difficult to place for adoption. Their intent was to optimize child health, but the series of legal cases set in motion to achieve this end necessarily involved involuntary control of pregnant women's conduct.

Liberal governance contains a dividing practice separating the free from the unfree. The practices of liberal governance have been systemically dependent on processes of disciplinary normalization that constitute citizen-subjects as capable of governing themselves according to a regulated freedom (Cruikshank 1999; Hunter 1994; Rose 1999). Those judged incapable of governing themselves in freedom have been and are governed through a variety of authoritarian means such as imprisonment and colonial rule (Hindess 2001). Those subjects – the colonized, children, women, the mad, and so on – who have not been governed through freedom provide opportunities for the growth of authoritarianism in biopolitics, the stuff of perfect cases, ideal exceptions for liberalism as the dissolution of freedom would there be of little notice. Thus it should come as no surprise that four of the seven reported cases brought by child protection authorities for the purposes of expanding their jurisdiction to contain the fetus involved Aboriginal women, with the remaining three being unemployed women of no fixed address. Aboriginal women were positioned as a pool of perinatal risk in the problematization of the threshold of the living subject. If liberalism contains a dividing practice separating the free from the unfree according to whether or not a collectivity or a person is considered capable of conducting themselves in accordance with a regulated freedom, Aboriginal peoples and those fallen from the working class have a long history of being classified as unfree and directly governed. In common with other colonial states (Hindess 2001), the relation of the Canadian federation (1867–) to Aboriginal peoples has been to confine, civilize and assimilate rather than to govern liberally. The tenuous relation of Aboriginal peoples to liberal governance put them in a position of vulnerability with respect to authoritarian projects in biopolitics.

In the line of child welfare cases discussed here, governmental styles of thinking attempted to assimilate legal reasoning, but failed on appeal. The site of the research thus bears on the question of the range of possible articulations between law and governance, that is, the various ways in which law and governance may be hinged. Historians of the present currently understand the normative power of law to divide the permitted from the forbidden, and to punish forbidden conduct, as having come to be systemically dependent on expert knowledges, operating in part through discipline and security (Ewald 1990; Hunt and Wickham 1994; Rose and Valverde 1998). In the problematization of the threshold of the living subject, biomedical styles of thought attempted to colonize law, suggesting that the biomedical concept of a living subject conserved before, during and after birth be fused with the concept of the legal person beginning after birth. The proposed conflation of the biomedical subject with the legal subject paralleled proposals found during World War II and the Nazi period to reduce the sovereign people to the demographic population (Agamben 2002: 84; Singer and Weir forthcoming). The articulation of law and governance proposed in these biopolitical projects was not one of articulating separate powers, but of assimilating law to governance. The end of the problematization of the perinatal threshold at law occurred with decisions of the Supreme Court of Canada ultimately confirming the birth threshold, the liberal governance of women and, by implication, the distinction between sovereignty and governance.

The child in need of protection

From the early 1980s child welfare agencies in several Canadian provinces and territories went to the courts trying to find some legal means to protect the unborn from health risks thought to be caused by maternal conduct during pregnancy. These moves by Canadian child welfare agencies paralleled the better known US legal decisions imprisoning pregnant women with addictions in the late 1980s and 1990s, the cases involving the so-called "crack babies", that is, women using crack cocaine during pregnancy (Campbell 1999; Daniels 1997; Litt and McNeil 1997; Logan 1999). The US and Canadian cases were similar in generating considerable publicity in the press, with the mass media demonizing women who used substances as selfish "monster mothers" utterly lacking in maternal feeling (Gomez 1997: 1–20; Greaves *et al.* 2002: 13–37; Rutman 2000; Tsing 1990). In both countries women charged with offences were from impoverished, racialized strata, in the US, primarily African Americans (Bower 1995; Ikemoto 1992; Kolder, Gallagher and Parsons 1987; Roberts 1997), and in Canada, Aboriginal peoples.[4] Whereas in the US, judicial intervention for "fetal endangerment" was consistently carried out under criminal law (Daniels 1993; Gomez 1997; Strickland and Whicker 1995), no parallel use of the criminal law was attempted in Canada, where child welfare agencies proceeded under a mix of child welfare law, mental health law, *parens patriae* claims, and tort law. Canadian agencies sought to order that women be detained for treatment rather than, as in many of the US cases, criminalized and imprisoned. In both nations, however, the strategies of confinement inevitably resonated with their shared colonial histories: the conquest of Aboriginal peoples and the enslavement of African Americans.

The child welfare system in Canada was born in the age of light, soap and water, a creature of late nineteenth- and early twentieth-century concerns with urban crime and disorder (Valverde 1991). Child saving organizations dedicated themselves to decreasing the "criminal and outcast portions of the population" (Kelso quoted in Rutman and Jones 1981: 54) by targeting "neglected children". In the words of J.J. Kelso, head of the Children's Aid Society in Ontario from its founding in 1891 and later first Superintendent of Child and Family Services in

Ontario from 1893 to 1934, "It is wiser and less expensive to save children than to punish criminals" (ibid.: 58). The child welfare regime was founded on a normative agenda of placing neglected children in "Christian homes" where, liberated from bad language, rude behaviour, poor dress and uncleanliness, they would learn the heaven-sent virtues of morality, temperance and industry. Throughout its history, the child welfare regime has been heavily allied with the cultural value of propriety, locked in battle to save children from parental (chiefly maternal[5]) "dirt and disorder" (Swift 1995: 74–77; Kline 1993: 327).

Child protection in Canada falls under provincial jurisdiction.[6] Provincial statutes of the late nineteenth and early twentieth centuries gave the concepts of "child protection" and "child neglect" legal force, incorporating child welfare agencies with broad powers to act on behalf of a new legal subject, the "child in need of protection". The first child welfare legislation was passed in the province of Ontario, where the care of neglected children was made a public responsibility under the first *Children's Protection Act* (1888). The 1888 *Act* confirmed the previous inherent jurisdiction of the courts to place neglected children in industrial homes and refuges, and extended the court's powers to allow the placement of children in authorized homes (Rutman and Jones 1981: 29–30). Under the subsequent *Act for the Prevention of Cruelty and Better Protection of Children* (1893), the Children's Aid Society, a previously existing organization, was incorporated with a provincial mandate. The 1893 *Act* granted the Children's Aid Society broad powers to apprehend neglected children, including the power of removing children from their parents and acting as their legal guardians.

The grounds for declaring a child in need of protection were broader at the end of the twentieth century than they had been at the beginning. Whereas late nineteenth- and early twentieth-century child protection services primarily concerned child neglect and abandonment, from the 1960s their activities increasingly focused on an expanding category of abuse, initially physical abuse during the 1960s, and then sexual and emotional abuse during the 1970s and 1980s (Bala 1991: 105). The contemporary mandate of child welfare authorities is the protection of children from neglect, abuse and exploitation. In all provinces and territories, child protection services have statutory power to investigate abuse and neglect, protect children by providing guidance and counselling for families, remove and provide care for children from familial situations judged abusive or neglectful, and place children in foster care or adoptive families (Bala 1991: 117–122). The care of children suffering from severe physical and mental health conditions whose families cannot or will not care for them also falls to the child welfare regime.

The history of the "child in need of protection", the legal subject first instituted at the turn of the twentieth century, could be studied as a succession of expert techniques to identify and find that child. The abstract categories of "neglect" and "abuse" in provincial statutory law have been given particular meanings through the forms of expertise brought before the courts in child protection hearings. As Mitchell Dean (2002: 57) has argued, "abstract and universal freedoms protected by legal powers" – such as parental authority – "are given normative content by specialized knowledges that frame the forms of life found within formally non-political domains of civil society". The use of expert witnesses to provide evidence at child protection hearings joined specialized, disciplinary knowledges about child abuse and neglect to abstract legal categories such as "neglect" or "abuse".

Until the 1980s child welfare authorities had not sought to protect a child prior to birth. Although not explicitly defined in child welfare law, the cultural standard of the birth threshold was presupposed as the point at which a child came into existence. Birth was the lower limit for intervention on behalf of the "child in need of protection" as a legal subject; a child could not be apprehended during or prior to birth. The move on the part of child welfare agencies to seek court orders protecting fetuses had two preconditions: (1) the formation of the perinatal threshold in health governance; and (2) the imbrication of child welfare in risk governance. The perinatal threshold, as I have shown in previous chapters, constructed time before, during and shortly after birth as one continuous interval based on commonalities in the bodily substance of the fetus and newborn. In health care, a risk-based regime of perinatal care was invented as one of the main techniques for optimizing perinatal health. This risk-based regime of perinatal care was the basis for evidence about fetal health by expert witnesses in legal proceedings. The perinatal threshold, made knowable through a heterogeneous ensemble of risk-based judgements, destabilized the prior implicit standard of the birth threshold when applied in child welfare proceedings. Evidence by health care professionals combined risk, diagnosis and prognosis in an uneasy mixture of medical judgements characteristic of prenatal care during the last four decades of the twentieth century.

A general trend towards the use of risk-based forms of reasoning was found in child protection services during the 1980s and 1990s. The problematization of child battery during the 1960s and 1970s, succeeded by child sexual abuse in the 1980s, had opened child welfare practice to risk management as a way of intervening in situations where witnesses to acts of abuse often did not exist, other than the child whose youth and trauma were thought to make recall difficult and unreliable for judicial purposes. Expert evidence provided in the form of risk judgements enabled child welfare authorities to intervene in suspected cases of child abuse that lacked proof supplied by "credible" witnesses to harm. The presence of risk reasoning in attempts by child welfare authorities to extend their jurisdiction to the unborn thus formed part of the broader opening of child welfare adjudication and social work practice to risk techniques during the 1980s and 1990s (Parton 1996 and 1998).

Contemporary child welfare practices internationally are saturated with risk judgements. Robert Castel's work (1991) has been fundamental to the interpretation of this trend, which he characterized as a move from the identification and targeting of dangerousness to risk. Examining the use of computerized data bases that predictively cluster risk factors, Castel documented how this form of risk management precipitated visits from child protection workers in France, sometimes to homes completely lacking "negative social welfare histories". Castel argued that risk assessment has displaced "the personalized relationship between professional and client" (Castel 1991: 282). Professionals currently perform assessments by reading individual case files rather than by meeting clients, a practice that marks a "shift from presence to memory, from the gaze to the objective accumulation of facts" (ibid.). Nigel Parton's (1996 and 1998) comments on child welfare practices in Britain accord with Castel's earlier analysis. Advanced liberal forms of child welfare, Parton maintains, focus on assessing and administering flows of risk factors rather than responding to the needs of children. In attempting to prevent harm, child welfare workers orient to identifying combinations of risk factors, a practice that Parton argues represents a dissolution of the subject of welfare: the individual child.

My concern here is with the imbrication of law and risk, a problematic marginal to the work of Castel, Parton and others who have analysed the significance of risk assessment to the practice of social workers in client relations. What may appear as the dissolution of the client/patient as subject at the level of expertise–client relations does not occur at the level of the legal subject, the child in need of protection. Judicial orders are made on the basis of harm, neglect or abuse, or, *risk* of harm, neglect or abuse; but they are always made in the name of the child in need of protection as legal subject. There is no attenuation of the legal subject, but rather a recomposition of its modes of identification through changes in statutory law and expert evidence. The question thus becomes one of tracing risk effects on the legal subject.

Risk governance ran like a prairie fire through the child welfare regime during the late twentieth century, attaching to both children and fetuses. While it is beyond the scope of this chapter to provide a history of risk governance in child welfare, a few observations may assist placing fetal risk in relation to the wider horizon of risk techniques deployed by child protection agencies during the 1980s and 1990s. Standardized, province-wide risk assessment regimes for child protection were put into place in the Canadian provinces of British Columbia (1996) and Ontario (1997) following inquiries into cases of child deaths where the children had been receiving services from child protection authorities (BCMCF 1996; Greaves *et al.* 2002: 52–53; Schmidt 1997; Chen 2003: 226–229). The province-wide risk assessment models had been preceded by, and were a response to, local and less systematic risk assessment by social workers during the 1980s and earlier 1990s. For social workers the requirements of formal risk assessment protocols obligated documenting and managing risk factors, decreasing time available for direct service work (Chen 2000).

The category of "risk" was explicitly present in the language of provincial and territorial child and family statutes prior to the standardization of the "Risk Assessment Models" during the latter half of the 1990s. The presence of "risk" in statutory law lent authority to preventing the harm, exploitation or abuse of children before it occurred. In Ontario, "risk" was first included in the identifying criteria for the "child in need of protection" under the *Child and Family Services Act, 1984*. Specific mention of "risk" was entirely absent from the definition of the "child in need of protection" in previous legislation, the *Child Welfare Act* (1980).[7] The 1984 *Act* had an iterative structure, beginning with three basic criteria for the child in need of protection. Such a child had suffered "physical harm", been "sexually molested or sexual exploited", or been subjected to "emotional harm" caused by the person in charge of him/her or by failure of that person to act appropriately according to the terms of the *Act*. These criteria were followed by three successive clauses broadening the definition of the child in need of protection to include: a child at "substantial risk" of "physical harm", "sexual molestation or sexual exploitation", or "emotional harm" from the person in charge of the child. A child might also be in need of protection as a result of the failure of the person in charge to act preventively with respect to the "substantial risk" of physical, sexual or emotional harm.[8]

When risk governance is enforced at law, judicial orders may be made in anticipation of illegal acts that have not occurred. Illustratively, one can point to a protection hearing brought by an Ontario child welfare agency under the *Child and Family Services Act, 1984*; the agency claimed that a woman's partner, who had in the past been convicted of assaulting her, presented a "substantial risk" to her child, despite his never having harmed the child.[9] The Ontario Provincial Court decided that her partner did represent such a "substantial risk" and ordered the child to remain in temporary care until the application for permanent wardship had been concluded.

Constituting the unborn as child through risk expertise

The explicit inclusion of "risk" in the Yukon *Children's Act* (1984), became thc legal basis for the first reported case in the Canadian jurisdiction of a woman being ordered into supervision and treatment for the sake of her "fetus", *Joe v. Y.T. (Dir. of Fam. and Children's Services)* (1986).[10] The court order was made under the *Children's Act* (1984)[11]:

> Where the Director (of Family and Children's Services) has reasonable and probable grounds to believe and does believe that a *foetus* is being subjected to a serious *risk* of suffering from fetal alcohol syndrome or other congenital injury attributable to the pregnant woman subjecting herself during pregnancy to addictive or intoxicating substances, the Director may apply to a judge for an order requiring the woman to participate in such reasonable supervision or counselling as the order specifies in respect of her use of addictive or intoxicating substances.[12] (emphasis added)

The language of this child welfare statute was unusual in being cast in terms of preventable risk to the *fetus* rather than risk to the *child*. The Yukon Territory

Children's Act (1984) was unique in Canada both at the point it was passed and since that time in making specific provision for *fetal* risk.[13]

In *Joe,* the Deputy Territorial Court Judge declined to hear expert evidence from a public health nurse because he considered fetal health risks from alcohol consumption during pregnancy as common knowledge of which he could take judicial notice:

> I believe that a judge is entitled to know and act upon his or her knowledge that a woman who drinks alcohol excessively while pregnant creates a situation where there is a considerable likelihood that the baby, when born, will be damaged in one or more ways. The baby's physical structure may be damaged, and the baby's intellectual capacities may have been diminished. Alcohol is the most common addictive drug in our society, and the fact that women who abuse it while pregnant damage their unborn child in that way is in 1985 common knowledge in Canada ... Alice Joe's unborn child is presently being subjected to a *serious risk* of suffering from fetal alcohol syndrome.[14] (emphasis added)

Alice Joe, who was an Aboriginal woman, was ordered into treatment because her fetus was deemed at health risk from her alcohol use.

On appeal to the Yukon Territory Supreme Court, the use of judicial intervention to secure the fetus against health risks from maternal conduct was declared incompatible with the liberty interests of citizens guaranteed under constitutional law. Justice Maddison, in a decision rich with droll sarcasm, observed that "fetal alcohol syndrome" was not defined in the *Children's Act* nor had its meaning been clarified at the original hearing, the result being that:

> He (the Deputy Territorial Court Judge) found that there was *serious risk* of the appellant's fetus suffering from fetal alcohol syndrome and everyone (including counsel) left the courtroom not knowing what it was there was a *serious risk* of the fetus suffering.[15] (emphasis added)

Justice Maddison searched medical dictionaries and periodicals to discover the meaning of fetal alcohol syndrome, but could find no standard term of art. Counsel for child welfare suggested that Alice Joe would have been able to understand the meaning of "fetal alcohol syndrome" had she simply asked the court for clarification. Justice Maddison replied acerbically: "The citizen should not have his liberty infringed by legislation which uses terms the meaning of which cannot be ascertained except by attending court."[16] Subsection 134 (1) of the Yukon *Children's Act* was struck down by Justice Maddison on constitutional grounds as a violation of s. 7 of the Canadian *Charter of Rights and Freedoms*, a constitutional provision, which guarantees the right to life, liberty and security of the person and the right not to be deprived of these protections except in accordance with the principles of fundamental justice.

The first reported case in Canada where the unborn was declared a "child in need of protection", *Re Children's Aid Society for the District of Kenora*,[17] was decided in 1981, and, like the subsequent 1986 case *Joe,* involved "fetal alcohol syndrome". A child, J.L., was born to C.L. in a Kenora (northwestern Ontario) hospital, six months after the physician who had confirmed the pregnancy, a Dr Beveridge, had contacted child welfare officials about his concerns regarding C.L.'s alcohol intake. Dr Beveridge had stated in writing at that time that C.L. had twice previously given birth when intoxicated, and that her alcohol use was continuing in the current pregnancy. His opinion was that C.L.'s alcohol consumption was "abusing and harming the unborn child".[18] The child protection agency did not act to apprehend J.L. until after she was born, a delay likely due to the fact that, as Tateishi (1989: 117) remarked, "the Society considered that the apprehension of the fetus was beyond its statutory authority". The baby was treated at birth for "fetal alcohol syndrome" by Dr Beveridge, who later gave evidence to the court about C.L's. history of alcohol use during pregnancy and J.L.'s fetal alcohol syndrome. Soon after birth the child was examined by a second doctor, a pediatrician, who also gave evidence that, "[I]n his opinion, there was a 95% possibility that the child was suffering at birth from the fetal alcohol syndrome."[19] Justice Bradley of the Ontario Provincial Court, Family Division, accepted the evidence of the two physicians, concluding that they had demonstrated J.L. as suffering at birth from "fetal alcohol syndrome" that had been "wilfully inflicted" by J.L.'s mother, "who refused to seek help for her alcohol problems despite the entreaties of Dr Beveridge".[20]

Justice Bradley declared J.L. a "child in need of protection" both *prior to* and at birth, under the *Child Welfare Act* (1980)[21]:

> by reason of the physical abuse of the child by the mother in her excessive consumption of alcohol during pregnancy, which conduct endangered the health of J.L., and further, by her neglecting or refusing to obtain proper remedial care or treatment for the child's health, when it was recommended by a legally qualified medical practitioner.[22]

The *Kenora* judgement constituted a legal precedent authorizing the apprehension of the unborn.[23]

The perinatal threshold formed the discursive precondition for the *Kenora* decision, enabling a connection to be established between the unborn and newborn both temporally and at the level of bodily substance. Interestingly, the decision in *Kenora* was dense with indirect medical discourse[24] introduced through the evidence of the two physicians, but it made minimal use of risk judgements. The statutory authority for the decision, the 1980 Ontario *Child Welfare Act,* did not contain explicit reference to "risk" in s. 19, which defined the conditions under which a child could be deemed in need of protection. As well, the physicians had pronounced their opinions of "fetal alcohol syndrome" as a diagnosis after the child had been born, with risk applying to the mother's past conduct in relation to the unborn during pregnancy.

Kenora was decided eight years after the introduction of the term "fetal alcohol syndrome" (FAS) by Jones and Smith (1973). The enthusiasm for FAS, apparent in *Kenora,* was characteristic of its early reception in medicine and the popular press. By the late 1970s medical researchers had rendered FAS a difficult diagnosis, suggesting "fetal alcohol effects" (FAE) for those who did not meet all the conditions for FAS (Clarren and Smith 1978). During the two decades that followed, a variety of terms was proposed, including alcohol-related neurodevelopmental disorder (O'Malley and Nanson 2002), alcohol-related birth defects (Warren and Foudin 2001), and prenatal exposure to alcohol (Jacobson and Jacobson 2002). Instability in nomenclature indicated uncertainties about diagnostic criteria. At both a popular and a clinical level there has been a solid belief in the effects of *maternal* prenatal alcohol use on the fetus. The intergenerational effects of *paternal* exposure to alcohol and other potential toxins have been investigated, but have attracted little legal, popular or clinical interest (Armstrong 2003; Daniels 1997; McCormack 1999; Schroedel and Peretz 1994).[25]

One year after the *Kenora* decision, a similar case – *British Columbia (Superintendent of Family and Child Services) v. McDonald* (1982) – was heard in the British Columbia Provincial Court by Justice Collings, who was asked to declare a child in need of protection on the basis of a mother's prenatal drug use.[26] As the child had never been in her mother's care, the court was asked to decide whether the mother's methadone use during pregnancy constituted "child abuse" within the meaning of the statute. The mother, B.M., was an Aboriginal woman. She had been on methadone (a replacement for heroin) at the point her pregnancy was confirmed by a physician, who believed that if B.M. stopped taking methadone, she might jeopardize her future child's health.[27] At birth the baby, D.J., was diagnosed as having "neonatal withdrawal syndrome",[28] and remained in hospital undergoing treatment over the next four months. Provincial Court Justice Collings' decision hinged on the use of the present tense in the test for a "child in need of protection" under the *Family and Child Service Act* (1980), which read: "'in need of protection' means, in relation to a child, that he *is* ..." (emphasis added).[29] Justice Collings emphasized the constraints placed on him by the choice of present tense in the language of the *Act*:

> I note the word is "is". There has to be an actual, not a notional deprivation of care, and the parent must have shown an actual, not a notional inability to care ... The problem I have with this is simply that since the child has been in hospital all the time I can't make an order.[30]

Birth marked the entry of a child into the verbal tenses of child welfare law. Justice Collings viewed the legal subject protected by the statute as having come into existence at birth and thus only those actions dating from the time of its birth were relevant to the apprehension order. The application was denied on the basis of the birth threshold.

On appeal to the British Columbia Supreme Court, Justice Proudfoot reversed the decision of the lower court in *McDonald*, and ordered the baby apprehended. Whereas Justice Collings had applied the birth threshold to define the point at which a child can be considered abused or deprived of care, Justice Proudfoot used the perinatal threshold, construing abuse as capable of occurring prior to birth. She held that a child at its birth can be retrospectively judged to have been abused during pregnancy. After considering the evidence both of Dr Segal (one of B.M.'s physicians during her pregnancy) and of the physicians who had treated the baby for withdrawal, Justice Proudfoot determined that the baby was "born abused":

> From the evidence before me of the physical problems that a baby born drug-addicted has to endure, it would be incredible to come to any other conclusion than that a drug-addicted baby is born abused. *That abuse has occurred during the gestation period.* ... D.J. falls within the definition under s. 1 as "a child in need of protection", as would any other child born drug-addicted. The Provincial Court Judge seemed to relate the "is" in that section to the necessity of the child actually living in the home and events which might occur thereafter. There is no such necessity required; *the child is born having been abused.*[31] (emphasis added)

Justice Proudfoot assumed the perinatal interval as a common-sense basis for normative judgement, making no distinction in the passage above between medical evidence and legal reasoning in the construction of a continuous temporal interval crossing birth.

The decisions in *Kenora* and *McDonald* resulted in the apprehension of children *after* birth. The later cases of *Re Children's Aid Society of Belleville, Hastings County and T. et al* (1987),[32] and *Re Baby R* (1987)[33] involved another scenario: seeking to apprehend a child *before* birth. Each case involved a woman who had refused medical treatment during pregnancy. Unlike the situation in *Joe,* where there was statutory authority for apprehending a "*fetus*" deemed to be at risk, child welfare legislation in the provinces where these two cases were heard, Ontario and British Columbia, made no mention of the fetus. *Belleville* and *Baby R* revolved around the legal issue of whether the unborn considered as a "fetus" could be encompassed within the meaning of a "child in need of protection".

Belleville was the first Canadian decision that resulted in the unborn being declared a "child in need of protection". By court order the fetus was placed in temporary wardship for three months. At first hearing in the Family Division of the Ontario Provincial Court, Justice Kirkland determined that "there was sufficient case-law and statutory authority for the court to make an order that a child 'en ventre sa mère' is indeed a child who can be found in need of protection",[34] but declared there was insufficient evidence to determine whether "this particular child was one in need of protection".[35] At a second hearing two days later, evidence was presented by the respondent's partner and his sister, who testified that the respondent, Linda T., had abdominal discharge and pain. A physician and a public health nurse, neither of whom had examined Linda T., gave evidence that such discharge and pain might indicate an infection endangering fetal life. Justice Kirkland summarized the medical evidence:

> Considering the developmental condition of the child, as related in the evidence of the public heath nurse, the child would indeed be in a *serious risk situation* where there exists an abnormal discharge and abdominal pain. These could be symptoms of infection within the foetus itself. The doctor's evidence that I heard last Thursday night indicates an infection can lead to pneumonia. This in turn could thereby result in the death of the unborn child.[36] (emphasis added)

However, Linda T. had declined all medical assessment and treatment. This concerned the Court on the grounds she might be *risking the health* of her unborn child/fetus:

> She refuses to seek, maintain or accept any form of medical assistance which is clearly necessary for the delivery of the child, particularly where there is a fear that the child could be born in an unhealthy state or in *a situation where the child's life is at risk.*[37] (emphasis added)

The judgement in *Belleville* slides together "fetus", "unborn child" and "child", making no distinction between medical and legal reasoning, treating all three as "in a serious risk situation". The Court raised additional issues about the respondent's "state of mind" as she had no fixed residence and had spent a night with her partner in an underground parking lot. In addition to her medical non-compliance, a rhetorical cloud of dirt and disorder hovered round Linda T., both factors demonstrating to the Court her want of maternal character.

Belleville involved an application by child protection authorities for an order apprehending the unborn child under mental health and child welfare statutes. Justice Kirkland determined that Linda T.'s conduct indicated a possible "mental disorder" that might cause bodily harm to either herself or her child; applying the mental health statute, he ordered her to be assessed by a doctor. The application under child welfare law was more complicated because it raised the legal question as to whether "the unborn child of Linda T. is a child in need of protection pursuant to the provisions of the *Child and Family Services Act* (1984) (Ont.), c. 55".[38] Child protection services claimed that "the conduct of the mother has placed the child in a risk situation".[39] Justice Kirkland modelled the maternal–fetal relation on the maternal–child relation after birth as one of "care": "The evidence today focuses on the circumstance of the child while still in the obvious and direct care of the child's mother."[40] He determined that the respondent's unborn child was a child within the meaning of the *Child and Family Services Act, 1984*, which specifies that a child is in need of protection where there is "a substantial risk that the child will suffer physical harm" or "the child requires medical treatment to cure, prevent or alleviate physical harm or suffering" and the parent or person in charge does not do so.[41] Referencing these provisions, Justice Kirkland reasoned that:

> if there is a *possibility* that this child will not be born alive, or that the child, although born alive, would be born with certain health defects, the court would be in a position to find the child in need of protection.[42] (emphasis added)

Justice Kirkland treated the evidence as showing that the respondent, by refusing medical care during pregnancy and declining to seek medical assistance in anticipation of the birth, created a situation where "the child's life is at risk",[43] and he consequently determined that the unborn, being at risk to its health from maternal conduct, was a child in need of protection.

In *Belleville* risk reasoning was present within the judgement, attaching to its argument. The Court allowed medical evidence about risk to prove there was a "substantial risk of physical harm", one of the statutory tests for a child in need of protection. The medical concept of prenatal risk was assimilated to the statutory sense of risk, which in turn was equated with possibility. This chain of associations attached the unborn child to the child in need of protection, with the unborn child/fetus in *Belleville* judged as at risk due to the pregnant woman's refusal of medical care. The domain of legal risk was made capacious in *Belleville,* potentially implicating all pregnancies judged at risk where women refuse prenatal care or do not comply with medical orders. One might recall that, in Linda T.'s case, the crucial evidence regarding her abdominal discharge and pain was given by her partner and his sister. Linda T. had never been evaluated by a physician. So too, the issue of where Linda T. might be when her "unborn child" was a ward of the Children's Aid Society received no mention in the decision, which was not appealed.

Re Baby R in 1987–1988 marked the last time that child protection services attempted to apprehend the unborn as a "child in need of protection" under child welfare law. Similar to *Belleville, Baby R* involved a woman who had refused medical treatment during pregnancy. R. was an Aboriginal woman. While in labour at the Vancouver Grace Hospital, she initially declined a Caesarian section recommended by the attending obstetrician who objected to assisting a vaginal birth in footling breech presentation (feet first). Breech position had been constituted as a risk factor for perinatal mortality in the survey reports of the 1960s; the obstetrician's opinion was a risk-based judgement conventional in perinatal medicine. In response to R.'s first refusal of the surgery, the obstetrician telephoned child welfare authorities, informing them that, "[t]he baby would die or would be seriously or permanently injured if the operation was not done".[44]

The social worker sought and received approval for the apprehension order from the Superintendent of Child Welfare, arriving on the scene to apprehend the fetus with two carloads of personnel from the Emergency Response Team, including three police officers and two social workers.[45] Prior to the arrival of the Emergency Response Team, R. reversed her decision, acquiescing to the procedure (Dawson 1990; Diduck 1993; Maier 1989).[46] As Karl Marx was fond of saying, history begins in tragedy and repeats itself in farce.

At the hearing in the British Columbia Provincial Court, Justice Davis received medical evidence from Dr Zouves, the obstetrician who had attended R., regarding his reasons for recommending a Caesarian section. Relying on this medical evidence, Justice Davis concluded that, "[t]he purpose of the apprehension was to ensure proper medical attention for the baby".[47] He then extended the jurisdiction of child welfare authorities to the unborn during and close to birth under conditions where women refuse recommended medical treatment:

> Under those circumstances, namely, where the baby is at or so near term and birth is imminent, the failure to provide necessary medical attention to prevent death or serious injury is sufficient to allow the superintendent to invoke the procedure of apprehension.[48]

Having settled the point of law that the child in need of protection extended to the unborn "at or near term", the question as to whether the court should order Baby R. apprehended was decided on the conventional basis of the mother's history and current status as evaluated by social workers and psychiatric evidence. R.'s prior history with child protection authorities damaged her character in the eyes of the Court, with Justice Davis beginning his summary of the evidence with the gloss, "The mother's history with children is atrocious. This is her fifth child and she has had only one of them in her care for more than a month."[49] Baby R. was thus determined to be a child in need of protection under the *Family and Child Service Act* (1980) because his/her mother was unable to care for R. due to neglect, lack of compliance with assistance, and an "emotional and mental condition".[50]

The Provincial Court established a perinatal threshold for child welfare services in British Columbia, but the order was equivocal as to whether Baby R.'s apprehension occurred before or after birth. On appeal of the case to the British Columbia Supreme Court, Justice Macdonell clarified the apprehension as having been authorized to take place prior to birth.[51] He defined the main issue for the appellate court as being "whether the unborn child is a child within the meaning of the *Family and Child Service Act,* so as to give the superintendent jurisdiction to apprehend".[52] Looking to that *Act* for clarification of the meaning of "child", he found sole guidance in the following: "'child' means a person under 19 years old". The act provided no other guidance as to the legal meaning of "child" and thus Justice Macdonell inferred:

> There is no definition of "person", but logically a person under 19 must only include children that have been born to be a child "under 19 years old", as you do not acquire an age until after birth. The previous Acts authorized the superintendent to apprehend a child only under certain circumstances, such as "children under the age of seventeen ... found begging in the streets", "were habitually truant from school", or "were dwelling with a thief, drunkard or vagrant". Obviously an unborn child could not come within the class above referred to. Additionally, up until recently it was necessary to bring the child before the court.[53]

The *Family and Child Service Act* implicitly defined persons in terms of the birth threshold, which in Euroamerican cultures construes birth as the null point, the beginning of the count in computing a person's age. So too, the practices of begging, truancy, theft, drunkenness, and vagrancy had not previously been attributed to the unborn. When the threshold of the living subject is extended prior to birth, inferences based on the birth threshold, e.g. the truant unborn, may be used to make the extension appear ridiculous.

Between the initial order for guardianship of Baby R. in 1987 and the time when the case came before the appellate court in 1988, two relevant decisions had been rendered: *R. v. Sullivan* (in the British Columbia Court of Appeal)[54] and *Re F* (Court of Appeal, England).[55] Although *R. v. Sullivan* had been tried under criminal law, Justice Macdonell considered its historical overview helpful, citing the comment: "At common law the line of demarcation for a fetus to become a person was the requirement that it be completely extruded from its mother's body and be born alive."[56] He then proceeded to apply the ruling in the English case, *Re F*, which had involved an application by a local authority for wardship of an unborn child; the Court had been asked to order the respondent restrained and compelled to have a hospital birth. The English Court of Appeal held that the unborn had neither legal existence nor legal rights, nor could it be made a ward of the court without intervening in the actions of the woman bearing it.

The social worker who had apprehended Baby R. had made a fine distinction when speaking with the obstetrician who had contacted child welfare about R.'s refusal of surgery. The obstetrician was advised

> to do what was required medically for the child but that he was not consenting to any medical procedure to be performed on the mother.[57]

Performing a Caesarian section without operating on R. would have been a test of the surgeon's art. That impossible distinction uneasily revealed what had not been explicitly addressed in the line of fetal apprehension cases to that point: the location of the unborn in women's bodies.

The appellate decision in *Baby R* made no explicit reference to the *Charter of Rights and Freedoms*, perhaps because two judgements of the Supreme Court of Canada involving challenges to the criminal law on abortion were then pending;[58]

both cases involved constitutional arguments relating to claims of fetal personhood and the liberty rights of women. However, Justice Macdonnell noted the argument made by the one intervenor in the appeal, the Women's Legal Education and Action Fund (LEAF).[59] A feminist non-governmental organization established after the adoption of the *Canadian Charter of Rights and Freedoms* (1982), LEAF's activities have involved the defence and promotion of women's equality rights, with a special focus on constitutional questions. The effect of LEAF's intervention in the *Baby R* appeal was to initiate recognition within Canadian judicial discourse that women's bodies form the site of apprehending the unborn. Referencing LEAF's arguments, Justice Macdonnell commented:

> The ramifications of a prebirth apprehension are self-evident, but need to be said as the effect of authorizing an apprehension prebirth of necessity means controlling the body of the mother to complete and effectuate a custody order.[60]

The judgement went further to note that, although Baby R. was apprehended during labour, the precedent of extending child welfare's jurisdiction to include the unborn would potentially open the making of custody orders to earlier points in pregnancy,[61] a power "to interfere with the rights of women" that would require "specific legislation and anything less will not do".[62] The order for the apprehension of Baby R. was set aside.

While the perinatal threshold is precisely defined for the purposes of official mortality statistics as deaths between twenty-eight weeks of gestation and seven days after birth, this definition had little salience for child welfare authorities. When the perinatal threshold was introduced into law, its precise temporal boundaries in health were of little legal interest. Its legal significance lay in connecting the child with the unborn across the birth threshold. As Justice Macdonnell reasoned, if a legal precedent permitting a prebirth apprehension were granted, it could be applied incrementally backwards with no obvious limit. In US jurisdictions, the concept of "fetal viability" can be deployed as the lower limit of apprehension on the basis of *Roe v. Wade*, which granted individual states the power to prohibit abortions late in pregnancy. However, the concept of "fetal viability" has no parallel in Canadian law (Martin and Coleman 1995: 963 n. 67). The point of legal demarcation – when the perinatal threshold begins or ends – is difficult to define: a classic "slippery slope" from the perspective of legal classification previously based on the birth threshold.

The decision of the appellate court in *Baby R* reaffirmed the birth threshold at law, but the ruling applied only within the province of British Columbia. Child welfare authorities in Ontario continued their legal strategy of attempting to have the unborn declared a "child in need of protection " in *Re A.* (*in utero*) (1990).[63] The Ontario Unified Family Court was requested to declare an unborn child a "child in need of protection" on two separate grounds: the Ontario *Child and Family Services Act, 1984* and the judicial *parens patriae* power to protect a child in place of a parent. The 1990 judgement of Justice Steinberg in *Re A,* followed the ruling and reasoning of the *Baby R* appellate decision.

In *Re A* a child protection authority applied for wardship of an unborn child that the agency judged to be at health risk as the result of the pregnant woman's refusing prenatal care despite multiple risk factors. The case review of Dr W.T. Conner, Justice Steinberg put the case for the high-risk status of the pregnancy strongly:

> The *potential risks* to the mother and child are that she may go on and develop eclampsia which is a condition of convulsions occurring particularly through labour with a *risk of mortality* for both mother and child. The implication is that she should certainly have been followed up closely medically over this last part of pregnancy and she should definitely have her baby in an area where she can be supervised both in terms of the delivery and also should she develop the significant medical complications of eclampsia. This means she should have her baby in a Level II or Level III hospital which implies a hospital with significant medical obstetric and anaesthesia backup.[64] (emphasis added)

As with the other cases discussed above dealing with the apprehension of the unborn, in *Re A* medical evidence of urgent perinatal risk factors lent normative, scientific authority to the application by child welfare services. Yet the line of jurisprudence denying apprehensions of the unborn under child welfare law was quite clear by the time *Re A* was heard. Citing the decision in the *Baby R* appeal, its approval by the Supreme Court of Canada in *Daigle* [1988], and the earlier case of *Dehler v. Ottawa Civic Hospital* (1979), Justice Steinberg concluded that Baby A. (*in utero*) was not a "child" within the meaning of child welfare law.[65] He refused as well to use the Court's *parens patriae* powers:

> here the child is actually inside of the mother. It is, therefore, impossible in this case to take steps to protect the child without ultimately forcing the mother, under restraint if necessary, to undergo medical treatment and other processes, against her will. I believe that the *parens patriae* jurisdiction is just not broad enough to envisage the forcible confinement of a parent as a necessary incident of its exercise.[66]

Rhetorically the decision in *Re A* set up a tragic tension between an urgent need to protect fetal health against prenatal and natal risk factors and the necessary limits of legal power.

When child protection authorities attempted to enlarge their mandate so as to encompass the unborn, they destabilized the relation between law and health governance at the threshold of the living subject. The line of cases leading from *Kenora* to *Baby R* problematized the relations between the threshold of the "child in need of protection" as a legal subject and the threshold of the living subject in health governance, between the birth threshold and the perinatal threshold, and between legal and medical knowledge. The legal decisions in these cases contemplated the implications of displacing the birth threshold and authorizing a perinatal one. In support of the child welfare initiatives, medical evidence of perinatal risk was offered at initial hearings, a litany of fetal and child risks from alcohol, methadone, infection, eclampsia, methadone, breech presentation and so forth. Physicians and nurses advised the courts of first instance about the risks to fetal and child health from the "non-compliant" conduct of women who refused to govern themselves according to the association between fetal and child health. If their applications had been accepted, child welfare authorities would have been authorized to protect the unborn, with their interventions defined by a constantly changing standard based on medical knowledge of perinatal risk. The result would have been to make court orders protecting the unborn dependent on medical evidence of perinatal risk. By the time of the appeal in *Baby R,* the authoritarian dimensions of court orders protecting fetuses had become apparent: invasions of women's liberty rights.

But this does not explain how it came to pass that four of the seven in these series of cases brought by child welfare authorities seeking jurisdiction over the unborn involved Aboriginal women: Joe, McDonald, R., and as we will see, G. Authoritarianism in biopolitics here rested on a colonial history of confinement and illiberal governance. Internationally the clients of child welfare authorities are drawn from impoverished social strata. The poorest people in Canada during the late twentieth century were Aboriginal peoples as reflected in standard indicators of income, employment, health, and housing. In 1981, the time of the decision in *Kenora*, 4.6 per cent of Aboriginal children were estimated to be in agency care, as compared to slightly under 1 per cent of the general Canadian child population (Johnston 1983: 57).[67]

Canadian governmental authorities have with great persistence developed policies that sever intergenerational links among Aboriginal peoples, the effect of which has been to reduce the numbers of people with Aboriginal cultures and identifications. During the 1950s and 1960s, a generation of Aboriginal people was removed from kin on reserves and sent for education to residential schools, places of notorious brutality (Miller 1996; Milloy 1999). After the demise of the residential school system, the governmental site for disconnecting Aboriginal children from Aboriginal parents and collectivities shifted to child welfare, with a sharp increase in the number of children in care from the early 1950s to the mid-1960s, a process referred to as the "sixties scoop" (Johnston 1983: 23). These numbers were publicized during the early 1970s and became the basis for political organizing that led to the formation of a parallel regime of child welfare agencies under Aboriginal control (Gray-Withers 1997; Schmidt 1997; Warry 1991). The Royal Commission on Aboriginal Peoples (1996, Vol. 3: 26) estimated that in most Canadian provinces (with the exception of Quebec) during the early 1980s, roughly 70–90 per cent of Aboriginal children in care were placed in non-Aboriginal foster and adoption homes. Aboriginal peoples have consistently interpreted the effects of child welfare as collective, arising from a long history of colonialism (Kline 1993; Monture-Angus 1995). The over-representation of Aboriginal women in the child welfare applications seeking to encompass the unborn in the meaning of child in need of protection thus did not occur through random demographic coincidence but was coherent with governmental power/knowledge practices in the second half of the twentieth century.

Refusing to make orders protecting fetuses

In 1996 child protection services made one last-ditch attempt to broaden their jurisdiction to include the unborn in *Winnipeg Child and Family Services (Northwest Area) v. D.F.G.*[68] The application was made under tort law, *parens patriae* and mental health law rather than child welfare law. *Winnipeg* marked the first time that child protection services had used the tort of negligence in their attempts to extend their mandate to the protection of the unborn. In its Factum in the appeal to the Supreme Court of Canada, the child welfare authority argued that the respondent, G., owed a duty of care to her fetus to refrain from activities having "a significant risk of grave and irreparable damage to the health and welfare of the child to be born".[69] The remedy proposed for G.'s alleged negligence, a court injunction for her detention and medical treatment, had no history of use in the law, which has been centrally about claims for damages. The application under tort law would have overturned both the live-birth rule and maternal immunity in torts of negligence for prenatal injuries. The suspension of maternal immunity would have made mothers and pregnant women liable for negligent acts to the unborn child they had carried or were carrying. Where the live-birth rule required a child to be born in order for a legal action to be launched on his/her behalf, its overturning would have meant that a claim could be brought during a pregnancy on behalf of an unborn child. Taking apart both the live-birth and maternal immunity rules would have undermined the legal fiction of the child *en ventre sa mère,* converting the fiction to a legal person in tort law.

The application by child protection authorities in *Winnipeg* brought to a ruthless conclusion the tendency already inchoate in *Duval* (1972) to identify the legal fiction of the child *en ventre sa mère* with the realist, biomedical concept of the fetus. The courts were offered another legal argument for the displacement of the birth threshold by the perinatal threshold. Ultimately the problematization of the child *en ventre sa mère* as the fetus, with the attendant collapsing of legal reasoning into the categories of health governance, was decided by the Supreme Court of Canada in favour of the birth threshold and the distinction of legal discourse from health governance.

With the exception of *Tremblay v. Daigle*, which was in part tried under the Quebec *Civil Code* (see note 92, Chapter 4 for discussion), no political controversy surrounded the development of case law pertaining to the legal fiction of the child *en ventre sa mère* prior to the *Winnipeg* case. Until *Winnipeg*, common law torts had no bearing on voluntary or involuntary risk reduction by pregnant women with respect to their own bodies. *Winnipeg* was the subject of political contestation in the legal process, the press and among Aboriginal organizations. Where *Montreal Tramways* and *Duval* had no legal intervenors, *Winnipeg* had eleven: three governmental, four Christian, three feminist and one civil liberties organization. The case also attracted considerable interest in the mass media, with G. portrayed as self-indulgently risking her future child's health through her solvent use (Rutman 2000: 64–68). G.'s representation in the mass media conformed to two sets of stereotypes: Aboriginal people as addicts, and pregnant women with addictions as heartless fiends in need of punishment (Rutman 2000: 69–71). The Aboriginal press was silent during the litigation (Rutman 2000: 71), although a number of Aboriginal organizations made statements, notably a joint statement co-signed by Health Canada and the Aboriginal Nurses of Canada, the Assembly of First Nations Health Commission, the Native Physicians Association in Canada and the Inuit Women's Association (see Health Canada 1996). The Aboriginal co-signers were organizations of major importance to Aboriginal health advocacy. The Métis Women of Manitoba and the Native Women's Transition Centre joined the Women's Health Clinic (a senior feminist health organization in Winnipeg) and the Manitoba Association of Rights and Liberties in a coalition that was granted intervenor status in *Winnipeg*. The above named Aboriginal organizations supported voluntary and "community-based" treatment options rather than judicial intervention in pregnancy. There was, however, no single Aboriginal position. A minority of Aboriginal organizations supported the legal tactics of child welfare in seeking judicial intervention and mandatory treatment for substance use during pregnancy. Across these differences in Aboriginal political and legal positions a common concern was shared for the negative impact of solvents, alcohol and other substances on Aboriginal health collectively.

G., the respondent/defendant in *Winnipeg*, had been orphaned as a young girl and placed in a number of foster homes. By the time G. had reached sixteen, she was living on the street and sniffing solvents. Her solvent use continued during her first three pregnancies (1990–1994); all three of her children had been removed from her care under permanent guardianship orders.[70] At the time of her third pregnancy, G. asked to enter a youth treatment programme at the Sagkeeng Treatment Centre, a First Nations facility on a reserve. She was denied entry because, at nineteen years of age she was no longer considered a "youth". During her fourth pregnancy G. requested treatment for her solvent use, applying to the St Norbert Foundation in Winnipeg, an alcohol and substance use treatment centre, but was told there was a waiting list of several months. In June 1996 child welfare authorities were informed of G.'s pregnancy and solvent use, responding by reopening her file and locating her. G. indicated she was willing to have treatment. A social worker went with G. for a prenatal examination and arranged for her admission to the St Norbert Foundation. When the social worker arrived to bring G. to the Foundation, G. was intoxicated, saying that she was willing to get treatment but "not right now".[71] Winnipeg Child and Family Services

(Northwest Area) (hereinafter the Agency) then immediately applied for a court order to confine G. for treatment.

Three witnesses testified on behalf of the Agency at the initial hearing before the Manitoba Court of Queen's Bench. Two were G.'s sisters, and the third the Co-ordinator of Services at the St Norbert Foundation. Five affidavits were filed in support of the Agency's position by Dr Laurie E. Hoeschen (Medical Director of the Chemical Withdrawal Unit at the Health Sciences Centre), Dr Albert Chudley (Head of the Section of Genetics and Metabolism at Health Sciences Centre), Dr Charles Ferguson (Director of the Child Protection Centre at the Health Sciences Centre), Kimberly Hansen and Marion Clement, these last two being G.'s social workers. Dr Hoeschen's affidavit outlined the health effects of solvents on users. Dr Chudley sketched the effects of maternal alcohol and solvent use on the fetus. Dr Ferguson reviewed G.'s hospital records for the birth of her second and third children. Kimberly Hansen and Marion Clement described their observations about G.'s solvent use. To these sources were added three reports on behalf of the Agency, one medical and two psychiatric. The medical report summarized G.'s history with the Chemical Withdrawal Unit. In the opinion of both psychiatrists G. was not suffering from a mental disorder within the meaning of the *Mental Health Act* (1987).

Justice Schulman of the Manitoba Court of Queen's Bench determined that G. was "mentally incompetent" under provincial mental health law, and that, unable to care for herself, she fell under the Court's *parens patriae* powers. He did not consider the application under the tort of negligence. G. was ordered into the custody of the Director of Child and Family Services, who was authorized to arrange her treatment, with the order terminating when G. had given birth. If G. failed to comply, child welfare was directed to apply without notice for an order committing her for treatment. Ostensibly the order was made on behalf of G.'s health, but the bulk of the reasoning and the medical evidence was about fetal health, with the exception of psychiatric evidence about G.'s mental competence. Although the order was stayed two days later, G. elected to stay on at the Health Sciences Centre until her course of treatment was finished four months later, almost three months after the Manitoba Court of Appeal had reversed the order.[72]

The decision of the Manitoba Court of Queen's Bench was reversed by the Manitoba Court of Appeal, where Justice Twaddle rejected the finding of mental disorder on the ground that the Agency had been concerned about the health of G.'s unborn child, not G.'s mental health.[73] The judgement of the Manitoba Court of Appeal set aside the order made under the *Mental Health Act* and *parens patriae*, noting that the two psychiatrists had not declared G. "mentally incompetent", and that the *parens patriae* of the Court could only be used to benefit the person on whose behalf it was exercised, not her unborn child.[74] The use of *parens patriae* was indicated only "where a finding of incompetence is justifiably made and then only for the purpose of benefiting the patient."[75]

Winnipeg Child and Family Services made clear that its concerns in the G. case lay in finding some means to secure fetal health in order to lower the incidence of childhood disabilities among those in its care. Two of G.'s children were disabled and had been made permanent wards of the state.[76] The Agency's Factum in the Supreme Court of Canada stated that, one year prior to *Winnipeg*:

> In January, 1995 Winnipeg Child and Family Services reviewed the files of the permanent wards in its care. At that time 146 out of 1076 or 14 per cent of the permanent wards had a medical diagnosis of F.A.S. or F.A.E.[77]

When children are declared permanent wards of the state they are no longer considered "adoptable" by child welfare authorities; they are likely to remain a permanent expense to the state, although they do remain legally adoptable. For the Agency, FAS/FAE constituted an administrative and budgetary problem that could be prevented if the Agency's mandate were adjusted to include fetal health. Later in its Factum, the Agency spoke of the need "to protect the child while still in the fetal stage".[78] Child protection in the fetal stage would require the replacement of the birth threshold by a perinatal one at law.

The Agency's figures were based on a conjunction of administrative records and medical knowledge. Child protection services used the medical categories of fetal alcohol syndrome/effects, identified instances of these in their case files, and then computed the disability rate among its permanent wards. The computation was a measure of prevalence: the number of cases of a disease of interest in a population at a moment in time. In terms of public health epidemiology, child welfare had identified those with an exposure (alcohol) and an outcome (FAS/FAE) but not those with an exposure and *without* the outcome. Whether the rate of FAS/FAE was increasing, stable or decreasing cannot be determined from these figures, which do not measure risk. Child welfare had used a prevalence measure as a scare tactic.

The social work evidence that G. was using solvents while pregnant and the medical evidence that solvent use puts fetal and child health at risk were linked to a legal argument that G. owed a duty of care to her unborn child under tort law. The use of risk reasoning to connect proof of fact in evidence with the legal concept, duty of care was at its clearest in the Factum of the Agency in the Supreme Court of Canada. Where *Montreal Tramways* and *Duval* had permitted actions in tort against third parties who had injured a woman while pregnant, child welfare authorities in *Winnipeg* proposed that pregnant women themselves had a duty of care to the unborn child they were carrying. Winnipeg Child and Family Services recommended this legal expansion on policy grounds:

> As a matter of policy, it is submitted that the duty of care to the fetus should apply to the birth mother as well as third parties. From the perspective of the child, it makes no difference if the harm done *in utero* was done by a third party or by the mother. As a society we should protect the most vulnerable.[79]

Although legal actions in tort typically seek damages, that is, people sue in the hopes of obtaining money or property for losses or harm they have endured, child protection authorities here sought to prevent harm before it happened:

> Damages are a poor substitute if the harm can be prevented in the first place. A remedy that seeks to prevent the harm before it occurs would be a much more preferable remedy than damages.[80]

Those who would not voluntarily comply with preventing harm before it occurred should, the Agency argued, be forced to do so. In the case of G., "The only way the agency could assist the child, that the mother had decided would be born, was to force the mother to abstain from the abuse of solvents."[81] Women whose conduct during pregnancy places their future child at risk should, it was reasoned, be compelled to abstain from their risk behaviours:

> The argument in favour of forced abstinence is simple. Preventing the ingestion of intoxicating substances substantially reduces the *risk* for the child being born with permanent defects that will affect the child's quality of life.[82] (emphasis added)

The Agency wanted sufficient powers to intervene where a pregnant woman's conduct placed the unborn/fetus at serious rather than moderate or minor risk:

> There are many activities that may be contra-indicated during pregnancy (poor nutrition, smoking) but the only activities that Appellant seeks to prevent are those where there is evidence that will satisfy a court that there is a *significant risk* of grave and irreparable damage that will impact on the quality of life of the child.[83] (emphasis added)

In effect child welfare authorities were proposing a risk standard of conduct taken from health governance for pregnant women .

The application by child protection services hinged not on proving that G.'s solvent use had actually harmed the health of the particular fetus she was carrying, but on something much more abstract: that fetal exposure to solvent use on a population basis has negative health outcomes for children after birth. After birth G.'s child showed none of the early signs of health effects from solvent use during pregnancy, to which it might be replied that the child welfare authorities had sought to prevent perinatal risk even in the absence of individual harm in order to reduce the population of children who were its permanent wards. This is the logic of population power in a foucauldian sense: the identification of aleatory events at a population level through abstractions such as mortality and life expectancy. In child welfare law, however, interventions do not occur through securing a population, e.g. through urban planning or purification of the milk supply, but through the application of population-level risk categories to individuals. Through a category error population power has become disciplinary, attached to the bodies of individuals.

The stakes in *Winnipeg* were high, proposing to take apart the principle of maternal immunity in tort law that prevented women from being held liable for prenatal injuries to the unborn child. *Winnipeg* and the subsequent case of *Dobson v. Dobson* (discussed below) threatened to undermine the fiction of the child *en ventre sa mère* and directly govern the health of the unborn through risk, compelling pregnant women – not only third parties – to imagine their bodies containing a class of subjects owed a duty of care that they could reasonably foresee. All conduct deemed harmful to the fetus/unborn would become potentially actionable, such as "pre-conception" torts involving maternal conduct prior to conception (Beal 1984; Shaw 1984; Weiler and Catton 1976) and "wrongful life" torts when pregnant women do not abort after they receive positive results from prenatal tests. The decisions of the Supreme Court of Canada affirmed maternal immunity, judging the proposals for its revocation as a fundamental challenge to the principles of reasoning found in common law torts.

Feminist intervenors in *Winnipeg* strongly supported the maternal immunity rule. The Women's Legal Education and Action Fund (LEAF) made a sharp distinction between third-party suits and suits taken out against women for their behaviour during pregnancy:

> Suits taken by children against third parties for damage caused *in utero* reinforce a woman's bodily integrity by providing an additional deterrent to negligent intrusions on her body and should not be used as precedent for obligations which would detract from that integrity and infringe her equality.[84]

The maternal immunity rule was also supported by the Canadian Abortion Rights Action League in its role as an intervenor in both *Winnipeg* and *Dobson*. In these cases, Canadian feminist organizations challenged neither actions for prenatal injuries nor the legal fiction of the child *en ventre sa mère*.

The Supreme Court of Canada in *Winnipeg* was divided, issuing a majority judgement and a dissent that displayed two quite different solutions to the problematization of the threshold of the living subject at law. The perinatal threshold had been articulated to law via medical evidence, particularly evidence of perinatal risk. The majority decision affirmed the birth threshold, disconnected legal and medical knowledge, and detached the legal concept of the child *en ventre sa mère* from the biomedical concept of the fetus. Asserting the distinction between law and health governance thus had the effect of exteriorizing the perinatal threshold from law, together with its associated logic of interventions based on perinatal risk. That the unborn were placed at health risk from particular conduct was thus not relevant to legal reasoning, although it remained relevant to health care. The Supreme Court rejected the proposal to make perinatal risk into a standard of conduct during pregnancy. The dissenting judgement proposed an alternative solution to the problematization of the threshold of the living subject at law: a perinatal threshold with a risk regime for pregnant women. In *Winnipeg* the dissent harmonized tort law with the perinatal threshold by reading the born-alive rule as rendered obsolete by contemporary medical knowledge. It argued in favour of a duty of care to the fetus/unborn, with pregnant women held to a perinatal risk standard of conduct knowable through medical evidence. As the legal fiction of the child *en ventre sa mère* was extended in the dissent to encompass the unborn/fetus, its status as fiction collapsed under the reality effects of biomedical discourse.

The majority decision of the Supreme Court of Canada in *Winnipeg* analysed the proposal to grant a tort-based injunction against a pregnant woman for the sake of her unborn child/fetus as involving fundamental changes to the law of tort. The Court understood itself as being asked to:

1 overturn the rule that rights accrue to a person only at birth (the "live-birth rule");
2 recognize a fetal right to sue the mother carrying the fetus;
3 recognize a cause of action for lifestyle choices which may adversely affect others;
4 recognize an injunctive remedy which deprives a defendant of important liberties, including her involuntary confinement.[85]

The majority decision concluded that, overall, the proposed extension of tort law was "of such magnitude, consequence, and policy difficulty"[86] that it belonged to the legislature, not the courts.

At issue in *Winnipeg* for the majority of the Supreme Court was the intelligibility of tort law. The decision turned first to examine the implications of the appeal for the legal concept of the "unborn child" in common law torts. The born-alive rule vests rights on birth. Thus, as we saw in the previous chapter, children born after the death of a parent have a right to claim inheritances or survivor benefits under statutory law and insurance plans. Overturning the born-alive rule would impose a duty of care to the unborn child that would "constitute a major departure from the common law as it has stood for decades".[87] The result would be to create legal rights before birth. Such recognition in turn would "reverse the long-standing principle of tort law that remedies for negligent behaviour cannot be pursued until a cause of action is brought by a juridical person".[88] The body during pregnancy would be the location of two persons. Justice McLachlin, speaking for the majority of the Court, noted:

> To permit an unborn child to sue its pregnant mother-to-be would introduce a radically new conception into the law; the unborn child and its mother as separate juristic persons in a mutually separable and antagonistic relation.[89]

Granting the unborn child a right to sue the woman bearing it would undo the construction of the "unborn child" as a legal fiction, literalizing it as a legal person, one in potential conflict of interest with its host.

The reality effects of medical evidence about the fetus that had been present in *Duval* (1972) were fully realized in the *Winnipeg* dissent. Justice Major, speaking for himself and Justice Sopinka, held that the increasing goodness of medical knowledge made the born-alive rule "outdated and indefensible".[90] The born-alive rule, maintained Justice Major, had been a rule of evidence rather than substantive law, a notion "rooted in rudimentary medical and scientific knowledge of the past".[91] Modern medical technologies give direct access to fetal reality:

> We no longer need to cling to an evidentiary presumption to the contrary when technologies like real time ultrasound, fetal heart monitors and fetoscopy can clearly show us that a foetus is alive and has been or will be injured by conduct of another. We can gauge fetal development with much more certainty than the common law presumed. How can the sophisticated micro-surgery that is now being performed on foetuses in utero be compatible with the "born alive" rule?[92]

The dissent of Justices Major and Sopinka treated the "unborn child" as a realistic representation referring to a world construed as anterior to discourse, a reality effect of medicine. Stripping the "unborn child" of its symbolic status as a legal fiction, the dissent equated the legal concept of "the child *en ventre sa mère*" with the biological concept of the fetus, collapsing legal into governmental reasoning.

The implications of negating the born-alive rule for the criminal law of abortion were discussed in the *Winnipeg* dissent. The Canadian criminal law of abortion had been struck down by the Supreme Court of Canada in *R. v. Morgentaler* (1988). Justice Sopinka reasoned that a woman has a duty of care to her unborn child only where she "has chosen to carry a foetus to term".[93] Under these conditions, the "woman must accept some responsibility for its well being".[94] In the dissent the duty of care to the "fetus" exists if and only if a woman has decided to continue a pregnancy.[95] Justice Sopinka proposed to classify the unborn child as a legal person owed a duty of care in order to assist "protecting the unborn child from having to live its life suffering from severe mental and physical disabilities."[96] The dissenting reasons maintain that this configuration of the duty of care meets the test established in *Daigle* (1989): "A foetus is treated as a person only where it is necessary to do so in order to protect its interests after it is born."[97] The logic of the reasoning in the *Winnipeg* dissent presupposed a perinatal threshold for its intelligibility: conserving fetal health for the sake of child health. Fetal–child continuity was posed in terms of *health* rather than *property* as originally envisaged in uses of the fiction *qui in utero est*, and as the Supreme Court of Canada had decided in *Daigle* with respect to the patrimonial interests of the "foetus" under the *Civil Code*.

The tort of negligence requires a juridical definition of a standard of conduct. Those who violate the standard can then be said to have failed in a duty of care. But what standard could be applied to conduct during pregnancy? How are pregnant women supposed to govern their own conduct with respect to what in this litigation is sometimes called the "unborn child" and sometimes the "fetus"? Justice McLachlin, speaking for the majority, noted that if the Court were to recognize a cause of action for a "lifestyle choice" harming others, the standard of conduct during pregnancy would not be clear, for "what will cause grave and irreparable damage to the fetus ... is a difficult endeavour with which medical researchers continually struggle".[98] Increased levels of scrutiny that necessarily involve uncertain and shifting standards would apply poorly to addictions, which "may be the products of circumstances and illness rather than free choice capable of effective deterrence by the legal sanction of tort".[99] The dissent in *Winnipeg* proposed an alternative standard for granting injunctions in tort against pregnant women and exercising *parens patriae* jurisdiction: where maternal conduct has "a reasonable probability of causing serious and irreparable damage to the fetus".[100] This articulates what might be called a "perinatal risk standard of conduct" for pregnant women. Estimates of the probability of negative health outcomes from exposures – risk estimates – would be based on current medical evidence, which pregnant women would presumably need to know in order to be held accountable for violations. The civil standard of proof was thus collapsed into medical estimates of risk that are presumed to saturate common culture. Justice McLachlin, speaking for the majority of the Court, noted that the burden of the new standard of conduct would fall disproportionately on "minority women, illiterate women, and women of limited education"[101] with the "difference between confinement and freedom" based on "a grasp of the latest research and its implications".[102]

The dissent in *Winnipeg* documented a wealth of medical evidence, the net effect of which was to constitute substance use during pregnancy as a national tragedy. Summarizing the facts of the case to a high level of detail, Justice Major provided twenty paragraphs outlining G.'s social welfare history, her medical history and the additional medical evidence presented to the Court of first instance. His summary of the facts contained detailed medical evidence about the effects of solvent use on the fetus that had been given by Drs Hoeschen and Chudley, as a preliminary to proposing what I have called a "perinatal risk" standard of conduct for pregnant women. The synopsis of Dr Chudley's evidence evokes the tone of terse horror found in Justice Major's reviews:

> Dr Albert Chudley, Head of the Section of Genetics and Metabolism at the Health Sciences Centre and a Professor of the Department of Pediatrics and Child Health and the Department of Human Genetics, stated in his affidavit that organic solvents used by chronic sniffers are neurotoxic to the brain of the foetus. Children exposed *in utero* to such substances may exhibit central nervous system dysfunction, developmental delay, attention deficit disorder

> and microcephaly. Experimental studies have shown that chronic exposure to solvents can lead to retardation of fetal growth and development and to fetal death. Dr Chudley testified that the critical period for the development of the central nervous system in humans is in the first sixteen weeks after conception. The central nervous system is, however, sensitive to a variety of toxic exposures until birth. Dr Chudley stated that any reduction of toxic exposure would reduce the central nervous system damage.[103]

The research referred to by Dr Chudley was not predictive at the individual level, hence the uses of "may exhibit" and "can lead" in this passage. A further fifteen paragraphs of "Additional Facts" (that is, in addition to the previous twenty paragraphs dealing with the "facts of the present case") reviewed current medical studies of fetal alcohol syndrome and fetal alcohol effects that had been presented in the Facta of Intervenors supporting the Agency's position, a complex dirge of relentlessly grim health news about the destruction of fetal and child health. On the basis of this medical evidence and G.'s social welfare history, Justice Major was of the view that G. should be ordered into treatment because her behaviour placed her fetus at risk of neurological and other forms of bodily harm.

Given the dissent's interpretation of the "born-alive" rule as a matter of evidence, and medical knowledge as increasingly perfect, Justices Major and Sopinka forged a close connection between legal and medical knowledge, with law as the enforcer of a standard of conduct constituted through medical risk techniques. In contrast, the majority judgement separated medical from legal knowledge, citing *Daigle:*

> The task of properly classifying a foetus in law and in science are different pursuits. Ascribing personhood to a foetus in law is a fundamentally normative task. It results in the recognition of rights and duties – a matter which falls outside the concerns of scientific classification. In short, this Court's task is a legal one.[104]

Canadian courts repeatedly insisted on the distinction between legal and medical knowledge in cases dealing with the government of pregnancy, an indication that attempts to fuse law with health governance were at issue in these cases.[105] In *Winnipeg*, this boundary exercise helped separate the legal fiction of the child *en ventre sa mère* from the biomedical concept of the foetus, resisting the reality effects of medical knowledge, including medical reasoning about risk. The disconnection between law and health governance prevented the creation of what I have called a perinatal risk standard of conduct for pregnant women.

Concluding her consideration of tort law, Justice McLachlin, speaking for the majority of the Court, sharply rejected the proposed extension of tort law to grant injunctive detentions: "the principles of tort law have never been used to justify the forcible detention and mandatory treatment of a person".[106] She declared the use of tort law to order a pregnant woman into detention to be a violation of liberty interests:

> The order at issue on this appeal can be upheld only by a radical extension of civil remedies into the most sacred sphere of personal liberty – the right of every person to live and move in freedom.[107]

The Court refused any extension of tort law for the purpose of detention, noting that these powers have been restricted to criminal law and orders made under provincial mental health law.

The Court was also sharply divided on the question of whether *parens patriae* extended to jurisdiction over the unborn child/fetus, with the majority adhering to the birth threshold and the dissent maintaining a perinatal threshold. *Parens patriae* has an old common law basis in the King's royal prerogative to become the guardian of those unable to represent their own legal interests (Ginn 1994). Its powers are equally awesome. Including the unborn child/fetus in *parens patriae* jurisdiction would have enabled child welfare authorities to apply for orders on its behalf such as medical treatment or place of residence. The majority decision denied the extension on the grounds that it would vest legal personhood in the unborn child, whereas "[t]he law sees birth as the necessary condition to legal personhood. The pregnant woman and her unborn child are one."[108] The effects of orders under *parens patriae* protecting the unborn child/fetus would "radically impinge on the fundamental liberties of the pregnant woman, both as to lifestyle choices and as to where she chooses to live and be."[109] The dissent preferred the approach of the lower court on the grounds that *parens patriae* was a broad power, and its extension appropriate so as to protect "the interests of those unable to protect themselves".[110]

The majority judgement in *Winnipeg* made repeated reference to fundamental liberties. The Court, however, did not base its judgement on constitutional law. The appeal failed because it was not supported by the legal framework of common law and *parens patriae*.

The problematization of the threshold of the living subject at play in *Winnipeg* had placed the child *en ventre sa mère* in jeopardy as a legal fiction. The majority and the dissent embodied two alternative responses to the problematization, with the majority adhering to the fiction and the dissent redefining it as a realistic representation. The instability present in the *Duval* decision, which on the one hand maintained the legal framework of the child *en ventre sa mère* as a fiction, and on the other treated the child *en ventre sa mère* as an entity that might be run over, was clearly split into two opposing judgements in *Winnipeg*. In adhering to a birth threshold for the living subject, the majority decision sustained the child *en ventre sa mère* as a legal fiction. The dissent reasoned otherwise, supporting a perinatal threshold, convinced that new medical knowledge permits seeing, hearing and operating on the fetus, thus rendering the fiction of the unborn child obsolete. In the dissent the child *en ventre sa mère* was literalized and law was conflated with health governance.

Legal fictions are not eternal. Fuller has argued that fictions have been removed from law by means of rejection or redefinition. Rejection occurs when a statute or judicial decision provides an alternative that discards the fiction. The meaning of the fiction is altered so that the "element of pretense" (Fuller 1967: 20) is no longer present; the figurative meaning is lost. In *Winnipeg* the literalization of a legal fiction, the child *en ventre sa mère*, through its redefinition as a biomedical fetus failed. The majority decision preserved the fiction, its utility linked to the intelligibility of common law torts and *parens patriae.*

With the decision in *Winnipeg,* child welfare authorities ceased trying to include the unborn child in their mandate, but others would continue the attempt to overturn maternal immunity in tort law. Two years after its judgement in *Winnipeg*, the Supreme Court of Canada heard another appeal, *Dobson v. Dobson*,[111] that sought to do this. Since both *Winnipeg* and *Dobson* involved the problematization of the threshold of the living subject in common law torts, some brief mention of *Dobson* is in order here.

The applicant in *Dobson* was the legal guardian of a child born with severe, permanent health problems after a Caesarian section at twenty-seven weeks gestation, an operation precipitated by an automobile accident in which his mother, Cynthia Dobson, had been a driver. The Supreme Court of Canada faced one main issue on appeal: whether to permit an action in common law torts against a mother whose child had been born with injuries allegedly caused by her negligence during pregnancy.

The decision of the Supreme Court of Canada in *Dobson* was much more strongly worded in terms of women's liberty and equality rights than *Winnipeg* had been. The majority of the Supreme Court again supported the rule of maternal immunity in torts, reasoning on the basis of women's privacy and autonomy rights together with social policy considerations:

> Although increased medical knowledge makes the consequences of certain behaviour more foreseeable, and facilitates the establishment of a causative link in negligence suits, public policy must also be considered. Significant policy concerns militate against the imposition of maternal tort liability for prenatal negligence. These relate primarily to (1) the privacy and autonomy rights of women and (2) the difficulties inherent in articulating a judicial standard of conduct for pregnant women.[112]

For the majority of the Court there was a "fundamental difference"[113] between imposing a duty of care to the unborn/fetus on a pregnant woman and imposing the same duty on third parties. In the latter case, a duty of care "does not significantly impair the right of third parties to control their own lives".[114] This is not the case in maternal–fetal relations: "Everything the pregnant woman does or fails to do may have a potentially detrimental impact on her foetus."[115] The result might potentially "render the most mundane decision taken in the course of her daily life as a pregnant woman subject to the scrutiny of the courts".[116] Taking apart maternal immunity would thus be highly invasive of privacy and autonomy as liberty rights. Nor did the majority think it possible to define a judicial standard of conduct for "prenatal negligence", or what the majority of the Court with dry wit (given the much-criticized androcentrism of "reasonable man" standards of conduct) called the "reasonable pregnant woman standard".[117] The application of a single standard would be vitiated by social variation in "financial situations, education, access to health services and ethnic backgrounds".[118] Any single standard would thus be arbitrary, and imposition of an arbitrary standard would violate the sphere of individual autonomy in conduct protected under civil liberties together with the "lifestyle choices"[119] protected in tort by the Court's decision in *Winnipeg.* Justices McLachlin and L'Heureux-Dubé, in concurring reasons supportive of the majority decision, applied a constitutional argument under the *Canadian Charter of Rights and Freedoms.* They maintained that the restrictions placed on pregnant women by dissolving maternal immunity in claims for prenatal injuries would infringe on women's liberty and equality rights under s. 7 of the *Charter of Rights and Freedoms.*

But what of Ryan Dobson born in the twenty-seventh week of pregnancy with a nasty case of cerebral palsy? The close proximity of his suffering to his mother's automobile insurance generated pathos, a cry for a certain kind of actuarial justice that would distribute the costs of caring for such children over the members of a risk pool. So blameless was this victim, so just his claim, that his mother's insurance carrier had voluntarily agreed to compensate his family. The Court suggested that, rather than setting a questionable precedent in common law torts, the needs of justice would be better served through specific legislation along the lines of the *Congenital Disabilities Act, 1976*[120] in the United Kingdom. This *Act* created an exemption from an exemption. It exempts women from liability for actions in tort brought by their children for injuries sustained prenatally. A child may, however, launch an action against his or her own mother for prenatal negligence if the harm was the result of the mother's negligent driving. The statutory exemption would be a much more limited intervention than setting a precedent under case law, as the latter would be subject to the unpredictable, piecemeal revisions of common law.

Complex cases at the articulation of law and health governance, *Winnipeg* and *Dobson* raised questions about how women were to be governed liberally. *Winnipeg* affirmed that the jurisdiction of child welfare authorities over the child in need of protection does not include the unborn. Thereafter, child protection services could recruit clients only from those who had crossed the birth threshold. The effect of *Winnipeg* was to decouple the perinatal threshold in health care from its extension into child welfare regimes where it would be used to order women into medical treatment for the sake of fetal and child health. The Supreme Court of Canada confirmed maternal immunity in common law torts, refusing to hold pregnant women liable for conduct that might place their "fetuses" at risk.

Since the time of *Winnipeg,* the treatment of pregnant women with substance use problems has relied on women's voluntary entry into programmes (Tait 2000). In the current situation, when child protection services become aware that a pregnant women is using substances thought to cause perinatal risk, they apply for a court order known as a "Birth Alert" that will result in the baby's apprehension at birth. Women are notified of the order and advised to enter a treatment programme, which, if completed successfully, may result in lifting of the order. Access to programmes is complicated by multiple barriers such as scarcity of programmes, long delays waiting for treatment, lack of daycare, programmes designed for men, military-style protocols based on an abstinence model, and a paucity of culturally appropriate facilities for Aboriginal women (Greaves *et al.* 2002; Rutman *et al.* 2000; Tait 2000). The standards of "successful" completion are not uniform, and non-compliance of any kind may result in a child apprehension. The will to perinatal risk reduction continues, now formally consistent with governing women in their freedom. This represents a change in risk management tactics with respect to fetal health to make them formally consistent with the liberal governance of women, not their defeat.

The series of legal decisions in the child welfare cases of the 1980s and 1990s attached risk reasoning to legal reasoning at three levels: in statements of fact, in the proposed extension of the legal person to encompass the unborn, and in proposals for a new standard of conduct. All were ultimately turned down on appeal.

References about risks to fetal and child health appeared when judges summarized the evidence of health expertise in statements of fact. In the cases decided under child welfare statutory law from *Joe* to *Re A,* much of the evidence about the health of the unborn took the form of risk estimates. This evidence was accepted in *Belleville, Joe, Kenora, McDonald* and the court of first instance in *Re Baby R* as proof that the unborn child had suffered physical abuse and neglect, and thus fell under the statutory tests of a child in need of protection. Risk evidence appeared most obviously in the factual summaries of these decisions, but it was also present in the legal analysis used to encompass the unborn child as a child in need of protection. In *Re A* and the appeal of *Baby R.* the same risk-based evidence was rejected as having no bearing on the legal meaning of the child in need of protection. In *Winnipeg*, the Supreme Court of Canada contemplated a change in the common law standard of conduct in tort law that would have taken apart maternal immunity. The revision would have held pregnant women to the standard of avoiding harm to the unborn children they carried, a standard that in the late twentieth century outside the courts was medically and culturally conceptualized through risk. In the proposal to take apart maternal immunity in tort law, the legal subject was posited as able to envision the consequences of her actions using a cultural repertory of knowledge based on perinatal risk. The argument for what I have called a "perinatal risk standard of conduct" was rejected by the Supreme Court of Canada.

In both the earlier litigation under child welfare statutes, and the later argued under tort law and *parens patriae*, risk reduction to the unborn child precipitated claims to the legal personhood of the unborn that were not explicitly argued by child welfare agencies. The arguments of child protection services came to be interpreted on appeal as justifications for encompassing the unborn within the meaning of legal person because the duty of care can only be to persons, and the concept of the child in need of protection applies only to legal persons. Risk governance of a biomedical subject (the fetus) invaded legal reasoning. This comprises an instance of what might be called sovereign risk, a catachresis to be sure, but at the threshold of the living subject powers become unstable, folding into each other.

Conclusion: the road from *Kenora* to *Winnipeg*

At the end of the twentieth century two thresholds of the living subject co-existed in differing forms of expertise. In health governance, a perinatal threshold had been instituted, stabilized in practices extending into kinship, the mass media and care of the self. The conclusion of the line of cases from *Kenora* to *Winnipeg* resulted in a reaffirmation of the birth threshold for the legal subject/person. In the Canadian jurisdiction, the perinatal threshold does not have legal force, but outside law its existence has proceeded apace.

These cases raise questions about the possible articulations of law and governance. Foucauldian conceptualizations of the relation between law and discipline/governance understand nineteenth- and twentieth-century law to be exercised through discipline/governance, with discipline/governance in turn stabilized through law (Ewald 1990; Hunt and Wickham 1994; Rose and Valverde 1998). Current foucauldian scholarship argues the mutual implication of law and governance, but also their exteriority, with law and governance being separate forms of expertise constituted through their own discourses, ends and subjects. Systems theory, in a way that differs markedly in analytic form from foucauldian scholarship, treats law as a system differentiated from other systems over the course of social evolution, with all subsystems being self-reproducing (Luhmann 1989). But what of instances where powers enter into competition, where one power seeks epistemic primacy over others, attempting to reconstruct the discursive practices of one power apparatus in terms of another? The differentiation of powers/systems is subject to challenge and contestation, as powers invade the space of other powers, even those, such as sovereign power, that are the necessary condition for the existence of other powers, such as governance.

In the line of cases that I have analysed here, child welfare authorities attempted to assimilate legal reasoning to health governance. These cases suggest that governance in our present does not recognize a boundary between itself and law, attempting rather to assimilate law to governance; these moments have been pervasively associated with biopolitics. Legal regimes have in turn found means for rejecting the encroachments of governance. The tensions between governance and law are of political significance since, during the twentieth century, governmental initiatives sought to elide the difference between sovereign people and demographic population, between legal person and biomedical fetus. The biopolitics of our present has been repeatedly given to such dangerous delusions which abrade the difference between the symbolic register of sovereignty that constitutes what it represents and the veridical discourses of governance that represent a reality posited as external to themselves (Singer and Weir forthcoming). It is through the symbolic powers of the democratic sovereign that the distinction between those who comprise the people and those who do not is constituted. A people does not exist in nature, nor does the legal person; they are symbolic rather than veridical constructs, unlike population and fetus. In order for historians of the present to better understand the articulation of sovereign power and governance, an urgent question with respect to biopolitics, it will be necessary to develop a more robust theoretical concept of sovereignty.[121]

The problematization of the threshold of the living subject initiated by child welfare authorities entered four forms of power into conflict: the security form of risk, disciplinary correction, the sovereign power of law, and liberal/authoritarian styles of governance. The security of the fetus was not legally compatible with the governance of women in freedom. It required confinement and mandatory medical treatment – disciplinary correction – that could only be accomplished by making an exception to women's civil rights during pregnancy and childbirth. The reception of the perinatal threshold at law instituted the notorious legal quandary of double personhood: fetus and pregnant woman with separate juridical rights. The temporal and bodily continuity of fetus and neonate which the concept of the perinatal crystallized in health governance was transformed at law into the impossible concept of two rights-bearing legal persons in one body. An alternative triangle of powers, discipline–law–authoritarian governance, was proposed to conserve fetal health in cases where security had proved inadequate. The legitimacy of this alternative strategy, what might be called governing through discipline, was at issue in the problematization of the threshold of the living subject. The law–security–discipline alliance proposed by child welfare authorities was turned down because the proposed change challenged the intelligibility of a legal regime that had been organized on the basis of the birth threshold.

The results of a problematization in biopolitics are not determined in advance. The power field of biopolitics at the threshold of the living subject is not a structure grinding out history as the repetition of a combinatory of underlying positions. The power field specifies what particular powers are inscribed in biopolitics together with their relations, but not the outcomes of political struggles. This was why waiting for the decision of the Supreme Court of Canada in *Winnipeg* was frightening for many disparate collectivities and subjects.

Freedom, law and security are not uniformly available to each and all; a text as foucauldian as mine perhaps needs to remember the obvious. Joe, McDonald,

R. and G. were Aboriginal women, that is, four of the seven cases brought by child welfare agencies for enlarging their jurisdiction to include fetal health ran through the lives of Aboriginal people. The vulnerability of Aboriginal women to their involuntary recruitment as defendants in these legal cases had to do with a long colonial history of authoritarian governance.

Only those collectivities and persons regarded as capable of governing themselves through freedom are granted the right of autonomy in liberal governance. Barry Hindess (2001) has emphasized that the history of liberal governance has been one of exclusions, one of these being colonial subjects. Aboriginal peoples have been part of these colonial histories, collectivities held incapable of liberal self-governance. With respect to Canada, this history of colonization has persistently deployed a repertoire of warfare, confinement and assimilationism against Aboriginal peoples, contested by Aboriginal demands for self-government and treaty relations with the Canadian state. The activities of child welfare authorities have been viewed as a form of assimilationism by Aboriginal peoples, a successor to the residential schools of the 1950s and 1960s in removing Aboriginal children from their kin and cultures. Most Aboriginal organizations and persons accordingly responded negatively to the child welfare litigation asking for court-ordered detention and mandatory treatment of pregnant women, placing it in the context of a long colonial history of coercion and assimilationism (Lundquist and Jackson 2000). In terms of the power field that constitutes the terrain of biopolitics at the disputed threshold of the living subject, Aboriginal women were weakly connected with the project of governing themselves and being governed through freedom. Their distance from the practices of liberal governance and proximity to the history of confinement made Aboriginal women ideal candidates for the purpose of setting legal precedents on the sovereignty–security–discipline axis of biopolitics. Risk governance acts on collectivities as well as persons, its course of action here following what Hegel called the slaughter bench of history.

6 Biopolitics at the threshold of the living subject

Care for the health of population, including its longevity, comprises one of the main functions of modern power, the others being war/peace, order and wealth (Foucault 1997d: 94). The genealogy of the health function constitutes the field of biopolitics within the history of the present. The investigations in which I have engaged here examined a key objective of twentieth-century biopolitics: optimizing health at the threshold of the living subject. In that century, care for the health of population strategically targeted the reduction of infant mortality over narrower and narrow intervals in the first year of life until the conservation of the living subject was pushed before birth. I have argued that the strategy of preserving health at the beginnings of the living subject led to the invention of a perinatal threshold, an alternative conception of the threshold of the living subject that unsettled the prior birth threshold. The resulting problematization of the threshold of the living subject in the second half of the twentieth century marked an event in vital politics that has not yet concluded.

In the preceding chapters I have taken up a set of feminist concerns characteristic of contemporary reproductive politics: reproductive freedom, the racializing of law and governance, the representation of the fetus as a subject with interests separable from those of women during pregnancy and childbirth. These feminist themes were read through the analytic conventions found in the history of the present. The result was to place the threshold of the living subject in relation to the fundamental strategic objective of biopolitics: conserving and optimizing population. But it was equally to consider biopolitics in relation to the threshold of the living subject. The analysis I have offered was thus doubled: feminist questions in relation to biopolitics, and biopolitics in relation to feminist inquiry.

The concept, threshold of the living subject, calls attention to the truth and symbolic regimes constituting the subject as it enters and leaves bodily substance. The living subject neither appears nor disappears instantaneously. The duration of that threshold and the character of the bodily substance across it have no ultimate normative definition secured for all time at the level of biology, medicine, law, or common culture. When the threshold of the living subject is unsettled, however, the effects bear on all those gathered there: the living subject, those who carry that subject, those who attend both these subjects, together with the multiple normative institutional practices that stabilize the threshold, notably law and kinship. The threshold of the living subject is inevitably thick with complex normative practices that come into question at the time of its problematization.

I have not undertaken a critique of the perinatal interval or the perinatal threshold as ideological concepts whose claims to truth must be unmasked as falsehoods. It was not necessary to postulate concealed interests that critique would reveal, since the work of power hid nothing while inventing the concept of the perinatal and fashioning techniques for the reduction of perinatal mortality. No doubt the works of power do sometimes proceed through obfuscation, lying and the presentation of partial interests as universal, but those considered in this book have proceeded in the full light of day, motivated in the name of health. They were announced in medical journals, policy documents and legal decisions. The dangers of the perinatal threshold lay with the effects of well publicized truths rather than hidden falsehoods.

The perinatal threshold was fashioned within medicine, formulated in response to the difficulties of lowering mortality in the first days of life after birth. As infant mortality rates fell during the first decades of the twentieth century, analysis of vital statistics showed that mortality close to birth remained intransigent. Through the efforts of physicians in pathological anatomy and social medicine, the causes of fetal death late in pregnancy and childbirth were demonstrated to be substantially the same as those in the first days after birth. The concept, "fetal and neonatal mortality", later renamed "perinatal mortality", measured deaths before, during, and shortly after birth on the basis of postulated commonalities in the bodily substance of newborn and fetus. The concept of the perinatal constructed a continuous temporal interval across the birth threshold, distinguishing the bodies of women from the fetus during the last trimester of pregnancy and the labour process. When perinatal mortality became reportable for the purposes of national and international systems of vital statistics during the late 1950s and 1960s, the reduction of the perinatal death rate became the standard against which both national health systems and maternal and child health services were judged. The conservation of population, each and all, had jumped the birth threshold, setting in motion new intensities of care for optimizing fetal and neonatal health.

The effect of the medical project to reduce mortality in the first days of life, a project pursued at its inception innocent of claims with respect to fetal personhood and devoid of religious conviction, was to constitute a new threshold for the living subject, a perinatal threshold, that unsettled the preexisting birth threshold. Under the birth threshold, the living subject was only definitely recognized at the conclusion of labour, that is, the point when the bodies of mother and baby were separated. Prior to birth, the existence of the living subject was recognized from quickening, but did not enter into definitive social recognition as a person and an individual until birth. The time between quickening and childbirth was clouded with uncertainty regarding the existence or non-existence of the living subject.

A threshold stabilizes a relation between inside and outside. It bears the between. The perinatal threshold unsettled the relation between inside and outside fixed by the birth threshold, moving the existence of the living subject before birth, and thus inside pregnancy and labour. The density of the truthful practices acting on the birth threshold, the very intensity of care acting so close to birth, displaced birth as the singular threshold of the living subject.

One might ask whether the displacement of the birth threshold simply annulled the threshold of the living subject. In what sense may one speak of the perinatal threshold as indeed a *threshold*?

The oldest meaning of threshold pertains to the timber or stone found at the base of a door to a dwelling. A threshold makes possible differentiation between the inside and outside of dwellings such that inside and outside are constituted as heterogeneous places with respect to each other. The main secondary meanings of threshold form figurative extensions of that ancient one: the border separating regions or fields, the beginning of an action, and a lower limit. In the early modern period, birth held apart interior and exterior, the hidden and the manifest, the *non-dum* and the child (Duden 2002a: 58–61). Birth was thus a threshold in the sense of making possible differentiation between the interior of a woman's body in pregnancy and the exteriority of an infant. After the invention of the biological fetus in the late eighteenth century, the epistemic quality of the *non-dum* as existent but hidden was clearly troubled within research medicine, but in clinical practice and common culture the birth threshold remained in place until the mid-twentieth century. An epistemic event occurred in the mid-twentieth century when the new medical concept of perinatal mortality posited a living subject in existence from the last trimester of pregnancy, across birth and into the first week. The labour of conserving the health of the fetus and newborn ran across the perinatal interval. The result was to institute a living subject that did not coincide with birth, a living subject prior to and during birth that was eventually determined not to be a legal person, although sometimes named and incorporated into kinship during pregnancy (Weir 1998b). Hence the perinatal threshold, if it is a threshold, does not coincide with the corporeal distinction between the interior and exterior of women's bodies at birth; the perinatal threshold is not organized on the spatial analogy of leaving a dwelling and crossing over to an outside. A biomedical construct, the perinatal threshold separated the threshold of the living subject from women in labour, undermining the previous significance of birth as the beginning point of the living subject, and thus potentially destabilizing the significance of birth as experience and ritual.

What then, if any, between is the perinatal threshold? The duration of the perinatal was instituted entirely in relation to securing health across an interval that had been characterized as having high mortality rates. The temporal interval between twenty-eight weeks fetal gestation, the birth process, and the end of the first week became a liminal period of dangerous trial both for health governance and for the fetus and newborn. The health of the living subject across the perinatal interval constituted a threshold in its figurative senses as a crossing over between regions, the commencement of an action, and a lower limit. The perinatal threshold is constituted on a population basis as a crossing over of living subjects from their non-existence prior to twenty-eight weeks to the assurance of their continued existence at the end of the first week. At the end of the perinatal threshold an upper limit is reached, passing over into another region; the peak of mortality just before, around and just after birth has been exited and the antepartum and intrapartum causes of mortality in the first year abated.

I have documented the rise of risk-based prenatal care as a technique formulated in the mid-1950s to reduce perinatal mortality and morbidity. A number of general analytic points with respect to risk governance follow from the genealogy of risk-based prenatal care and the midwifery orientation to risk. Risk techniques are analytically disparate, as illustrated in the discussion of epidemiological and clinical risk. In epidemiology, risk forms a quantitative measure calculated by examining a study population over a defined time interval to see how many new cases of a disease/condition appear among those who have had the exposure of interest to the research. Clinical risk moves the epidemiological calculations of risk factors to the diagnosis and care of patients, where population-based calculations are not prospectively predictive. Clinical risk governs through uncertainty. It combines differing forms of medical reasoning: epidemiological knowledge of risk, diagnoses, test results. An amalgamation of incommensurable judgements, clinical risk has been criticized within health expertise precisely for the effects of binding together risk with diagnostics, which results in risk assessment tools having an inflated reliability.

My work has pointed to the significance of investigating risk in relation to other styles of reasoning within the health complex that act to compete with and restrict risk reasoning. In the study of midwives as risk assessors, I have shown that midwives use clinical judgement of the normal and the pathological in an attempt to limit the use of risk management in routine pregnancy care. Midwifery is energized by a professional ethos oriented to pregnancy as a state of health, privileging clinical discourse and research-based medicine as a basis for care. Midwifery care of pregnancy in a state of health involves social, experiential and cultural relevances beyond those found in risk-based prenatal care which aims to reduce perinatal mortality and morbidity. Both medicine and midwifery as forms of health expertise possess their own internal critiques of risk techniques. They each contain styles of reasoning, notably the clinical judgement of the normal and the pathological, but also the quantitative techniques of clinical trials studied in evidence-based medicine, that act to contest risk reasoning in health care. The result is to maintain bases in the contemporary health sciences for normative judgement exceeding risk reasoning.

A wealth of social science studies has documented the uses of risk in neoliberal health governance. Risk techniques encourage individuals to be responsible for their health status by avoiding selected substances such as smoking and alcohol and loving exercise and green vegetables. While insightful, studies of neoliberal risk governance comprise a narrow research trajectory too much focused on a small number of risk effects. Routine risk assessment of pregnancy was not a neoliberal invention, nor has its primary impetus been located in the health promotion efforts of public health. An alternative and more satisfying genealogy of risk would begin in the period after World War II with the invention of analytic epidemiology ("risk factor" epidemiology). Many of the risk factors that have fascinated social scientists were first isolated in the cohort studies of chronic disease begun at that time, notably the Framingham study which continues to date (Dawber 1980). So too, the attention to risk in social science studies has been coupled with a marked disinterest in other forms of quantification in the health sciences and health care, especially clinical trials and the growth of evidence-based medicine which seeks to transform the bases of clinical judgement in our present (Mykhalovskiy and Weir 2004).

In the second half of this book, I examined the reception of the perinatal threshold at law when the courts were presented with medical evidence of a living subject knowable through risk reasoning prior to and during birth. This investigation linked a literature on risk governance with a separate and largely feminist literature on the legal address of pregnancy. My account traced perinatal risk effects on tort and child welfare law. Medical evidence of perinatal risk led to a problematizing of the legal classification that sustained a birth threshold at law, linking mortality and morbidity after birth to fetal risk. A line of child welfare cases proposed overturning the birth threshold, substituting a perinatal one in its place. These cases were fundamentally motivated by questions of health, specifically the need to prevent injury to fetal health later in pregnancy and labour in order to conserve child health. The concern with fetal *health* was distinct from other cases argued under Canadian constitutional and criminal law that sought to have the fetus declared a legal person on existential grounds. At law the unborn configured as a biological fetus has been mobilized to differing ends by diverse political interests.

In both the negligence torts and child welfare cases, the introduction of the perinatal threshold threatened to make women responsible for conduct during pregnancy that placed the health of their future children at risk. Taking a long historical view, it is astonishing that the reality effects of perinatal risk evidence destabilized the ancient legal fiction, *qui in utero est*/the child *en ventre sa mère*, which dated to the *Digest* of Justinian, by suggesting it was equivalent to the biomedical concept of the fetus. In the litigation brought by child welfare authorities attempting to enlarge their mandate to include some and any legal means of preserving fetal health in order to reduce the numbers of sick and disabled children in permanent care, medical evidence of perinatal risk was used as proof of fact to justify intervention. Epidemiological research on fetal risk and the clinical knowledge systematized in risk-based prenatal care was moved from its previous medical location into the courts, suggesting a perinatal threshold be applied in child welfare litigation and damage claims for prenatal injuries.

The effects of truth at the perinatal threshold were pursued with particular vigour in relation to Aboriginal women. Given the significance attributed to perinatal mortality as a measure of national health and health care, the higher comparative perinatal mortality rate among Aboriginal peoples became an object for governmental interventions after surveys during the 1970s showed that Aboriginal rates, particularly those in the North, were higher than that found in the general Canadian population. Aboriginal women during pregnancy came to be seen through the lens of perinatal risk from two directions: health governance and child welfare. Health policies authorized the removal of Aboriginal women late in pregnancy from Northern reserves to places of birth in Southern hospitals. In this process, previous Aboriginal cultures of childbirth were undermined. During the 1980s and 1990s when child welfare agencies launched legal actions trying to find some means of controlling the conduct of pregnant women that the agencies believed put the health of their unborn and future children at risk, four of the seven reported cases involved Aboriginal women. The perinatal threshold and its techniques of implementation with respect to Aboriginal peoples have thus been associated with aggressive neocolonial governance. The problematization of the threshold of the living subject was bound up in a long colonial history, the effects of truth racialized at the threshold of the Aboriginal living subject.

I have concentrated on risk as a technique for implanting health across the perinatal interval, studying it genealogically as a truthful practice. However, practices for conserving perinatal health comprise much more than risk techniques. To understand how the perinatal threshold operates more broadly other ranges of practice would need to be explored, such as perinatal care, perinatology, perinatal death reviews in hospitals, and perinatal risk measures for newborns. Ethnographic studies are needed to scrutinize patterns of social relations and interaction that fall outside the scope of genealogical investigations.

When the reduction of perinatal mortality was instituted as an objective of health governance, it effected profound transformations in the culture of pregnancy and childbirth. Perinatal risk factors were folded into the previous practices of prenatal care, resulting in a regime of higher-intensity care during pregnancy and labour. In the name of reducing perinatal risk, the fate of home birth, already weakened during the 1930s and 1940s, was sealed. Little reflection was given to the theory of living, that is, ethics in the Ancient Greek sense, that these forms of health governance presupposed until social movement challenges of the 1970s and 1980s. The costs of perinatal medicine to the living, including health care providers caught between governmental scrutiny and client expectations, fell outside the parameters of perinatal mortality surveys. As governance in maternal and child health services came to be driven by the goal of reducing perinatal mortality and morbidity rates, it incited the impossible popular hope of driving death from pregnancy and the birth process. In the Canadian North, some Aboriginal peoples responded to the airlifting of Aboriginal women for birth in Southern hospitals by saying that they would not live in the North if they wished to avoid risk, and that their cultures should include birth (Kaufert and O'Neil 1993). Contestations about appropriate birth technologies implicate normative decisions about ways of living. These broader questions about the costs of perinatal medicine for our present need further reflection.

I have been concerned to write the problematization of the threshold of the living subject in relation to its authoritarian implications for women. The perinatal threshold is, as I have emphasized, a temporal division folded into the maternal body during pregnancy and labour. Across the perinatal interval, fetus and newborn were rendered precious lives. Fetal demise and deaths of newborns were equally perinatal deaths. The perinatal threshold thus systemically generated maternal–fetal division by extending the duration of the living subject across birth and into the time of pregnancy. This analysis reads the emergence of the unborn/fetus as a living subject for governmental and clinical purposes in the context of the strategic objective to reduce the perinatal mortality rate. The appearance of maternal–fetal division is thus situated within a logic of population politics dating to the mid-twentieth century rather than to the more recent dissemination of visual imagery pertaining to the fetus such as prenatal ultrasound.

Failure to care for perinatal health antenatally and intranatally proves consequential for health care providers and for women during pregnancy. Obstetrical departments in hospitals with high perinatal mortality rates may be decertified. Canadian courts contemplated whether pregnant and birthing women who incurred risks to fetal health might be subject to court orders for compulsory risk reduction. Those who are judged incapable of governing themselves autonomously may be exempted from liberal freedoms and governed directly for their own good/the good of others. With respect to the biopolitics of the perinatal threshold, the move from freedom to unfreedom corresponds to risk compliance secured through two separate powers: security and discipline.

Biopolitics is not a single power but a field comprised of interacting, agonistic powers in strategic relations with each other, what I have called a power field. When the threshold of the living subject came to be problematized, the strategic relations among the powers mobilized at the threshold became unclear. The powers mobilized at the threshold of the living subject were linked together in two alternative strategic fields: (1) security–liberal governance–sovereign law; (2) discipline–illiberal governance–sovereign law. The sequence of security–liberal governance–sovereign law, what might be called governing through security, was applied on a population basis in prenatal risk assessment. In this strategy women were configured as liberal subjects. In the second of the strategic fields, governing through discipline, women's freedoms as liberal subjects were suspended. Legally ordered discipline was invoked where security had failed to incite voluntary adherence to risk-aversive conduct, precipitating attempts on the part of child welfare and health personnel to seek legal authorization to suspend women's freedoms and compel obedience to conserving fetal health. The two strategic configurations of the power field co-existed, although governing the perinatal threshold through disciplinary correction was later turned down at Canadian law.

Although I have investigated biopolitics at a particular site, the threshold of the living subject, these strategic patterns can reasonably be expected to run through biopolitics more generally. Those who fail to be incited by the promise of security are considered for disciplinary correction. The power field of biopolitics may, however, contain additional powers and strategic configurations not present during the problematization of the threshold of the living subject. These remain to be investigated, preferably on the basis of empirical research.

Power/knowledge relations at the perinatal threshold persistently lose their analytic specificity. The powers and techniques so carefully distinguished by historians of the present encroach on each other. Security and discipline hybridize in clinical risk; governance invades law; the risk–uncertainty relation becomes unstable.

Clinical risk management at the perinatal threshold projects the epidemiological/security analysis of populations to the disciplinary level of individual bodies. Foucault, however, suggested that security and discipline are mutually enforcing but analytically distinct. Security is a power/knowledge apparatus constituted from the analysis of aleatory events in large populations; regularities are conceptualized and identified through abstractions such as mortality or life expectancy, and mechanisms are devised to intervene in the abstracted events at a population level. Discipline acts to optimize and correct the performance of individual bodies. Crosscutting security and discipline, clinical risk is, as I have shown, an amalgamation of incommensurable forms of reasoning assembled under the rubric of risk. In the process of moving from population to individual, clinical risk enters uncertainty.

As to the governance–law relation, the problematization of the threshold of the living subject in child welfare litigation of the 1980s and 1990s proposed to collapse the sovereign power of law into health governance, suggesting that the child as legal person be extended to cover the unborn configured under the biological concept of the fetus. Child welfare agencies wished to have their mandate expanded to control maternal conduct in order to reduce fetal health risks for the benefit of child health. Although turned down on appeal, this legal strategy is noteworthy for its attempt to collapse the distinction between governance and sovereignty, one of the recurring dangers found in the biopolitics of our present.

Last, the relation between governing through risk and governing through uncertainty becomes indistinct at the perinatal threshold in both the health and the legal regimes. Clinical risk binds together risk judgements, test results and diagnoses in a motley heap. The result is incoherent, and even those judgements that might technically have an epidemiological basis fall in technical uncertainty at the individual level. At law, fetal health risks presented in expert evidence were argued to justify a new standard of conduct for pregnant women who, if it had been allowed, would have been reasonably expected to foresee the probable results of their conduct under conditions of uncertainty.

At the perinatal threshold the logic of security invades discipline; governance attempts to engulf sovereignty but is rebuffed, and risk continually dissolves into uncertainty. The utopian imagination is well and truly alive in biopolitics, if not elsewhere in advanced modernity, and it is given to desperate measures.

Authoritarianism (in the sense of a suspension/exception to governing through freedom) is intrinsic to conserving the health of population, that fundamental function of modern power. The forms of authoritarian practice have been many, ranging from quarantine measures in public health to the involuntary recruitment of subjects for medical research. The problematization of the threshold of the living subject in child welfare litigation surfaced at the site of one of the lasting conundrums of biopolitics: the subject-citizen who violates the duty to be well, the subject who will not freely govern him/herself according to health, instead refusing "life-giving" medications, blood transfusions, surgery, and so forth. Refusing health has been deeply puzzling to a biopolitics seeking to govern health through freedom. How can anyone not believe in health? What practices of health may be compelled? At the perinatal threshold the recurring dilemma of how to govern those who refuse or ignore the value of health was applied to pregnant and birthing women who did not orient their conduct to the security of fetal health when the tissues of the living subject were not yet separated from their generative matrix.

The problematization of the threshold of the living subject has been a transnational phenomenon present in those nation-states that have adopted the reduction of perinatal mortality and morbidity as a health governance objective. Yet the results of any problematization are contingent – their effects cannot be known in advance. In the Canadian jurisdiction, the birth threshold was affirmed in a series of legal decisions at the end of the twentieth century. Thus in Canada and other jurisdictions such as the United Kingdom, though notably not in the United States of America, two thresholds of the living subject currently co-exist: a birth threshold at law and a perinatal threshold in the health complex. In those nation-states that have reaffirmed the birth threshold at law, the perinatal threshold is still very much present in health governance, spreading from health into kinship and the mass media. The future of the perinatal threshold of the living subject in health, kinship and common culture has not yet been settled: a Northwest Passage, its effects obscure and massive.

Notes

1 On the threshold of the living subject

1 "*Die Schwelle trägt das Zwichen*" (Heidegger 1959: 26).

2 "*Er hält die Mitte, in der die Zwei, das Draussen und das Drinnen, einander durchgehen, aus*" (Heidegger 1959: 26).

3 The "perinatal subject" is absent because fetus and newborn have not been unified as a single subject in contemporary biomedicine.

4 The transition between the quick and the dead, what I have called the threshold of exit, has received more scholarly interest than the threshold of entry. See particularly the extraordinary ethnography of Margaret Lock (2002).

5 A third relation between bare life and form of life has now occurred: the cultural alteration of bare life, that is the fusion of bare life and form of life. Paul Rabinow (1996a) calls this the "biosocial". I do not discuss the question of the biosocial here as the cases of risk governance being investigated do not deal with cultural modifications of pregnancy at the level of bare life. Risk governance of pregnancy affects all women receiving prenatal care today, whereas the techniques of "new reproductive technologies" affect few. This is not to disqualify interest in biosocial modes of pregnancy on the grounds that it reflects the interests of white world elites rather than the maternal and child health needs of the majority of the women globally (Singh 1997). Of course this is the case, but it misses the culturally transformative potential of biosocial modes of pregnancy over the next decades. More to the point here is that the "new reproductive technologies" constitute an exception to the risk governance of pregnancy. The relation of "new reproductive technologies" to risk and its measures, for instance the ways in which *in vitro* fertilization is associated with increases in the perinatal mortality rate, deserves separate investigation.

6 "Assemblage" enters the history of the present via the work of Deleuze. An assemblage integrates a functional, strategic aim (e.g. education) with a visible form (e.g. the school), acting to produce a particular kind of human subject (e.g. the student) in a field of discourse (Deleuze 1988: 37–39). Mitchell Dean (1996: 55) notes that "assemblage" represents "a way of thinking about entities as multiplicities rather than as unities, as a complex ensemble of discontinuous elements and forces".

7 For overviews of how the history of the present reads liberalism, see Dean 1999a: 113–130; Gordon 1991: 14–46; Rose 1999: 61–97.

8 This is what Foucault (1988a: 113) calls the "polymorphism" of liberalism – the ways in which liberalism contains an internal capacity for reflection and reform, separating its present and future from past "abuses". Liberalism is prestructured for conflict.

9 The phrasing in this passage is intentionally suggestive rather than definitive. It extends a critique of the virtual disappearance of sovereignty in Foucauldian work, a critique undertaken with my colleague Brian Singer (Singer and Weir forthcoming), to

biopolitics. The claims being advanced are that sovereignty is implicated in contemporary biopolitics, that sovereignty does not conform to a logic of a truth regime, and thus that biopolitics is not wholly veridical.

10 For Deleuze (1988: 44), the "diagram" represents a way of characterizing historical epochs lasting centuries; the "diagram" is a mapping coextensive with the whole social field which enables a society to be globally characterized as "sovereign" or "disciplinary". Deleuze's reading would appear to be at variance with Foucault's comments on multiple types of power existing simultaneously.

11 "Epidemiological risk" and "clinical risk" are terms first used in Gifford's influential 1986 article.

12 On the "psychological complex", see Rose 1985: 9; for "legal complex" consult Rose and Valverde 1998: 542–3.

13 My earlier discussion of clinical risk (Weir 1996) was an attempt to think outside the box of actuarial risk. It thus used a conventional argumentative form, placing clinical risk as antithetical to actuarial risk. The argument has the disadvantage of implicitly presuming clinical risk to be the only form of pregnancy risk governance. The present study locates actuarial techniques within the governance of pregnancy rather than outside it, included among the risk forms examined here: epidemiological, clinical, actuarial and legal risk.

14 *Montreal Tramways Co. v. Léveillé*, [1933] S.C.R. 456, 41 C.R.C. 291, 4 D.L.R. 337.

15 The question of governance is separable from the medical and midwifery history of pregnancy and childbirth, having to do with medico-administrative rationales and techniques for the governance of population (the conduct of conduct) rather than health provision (conduct). In this sense, the history of care during pregnancy and childbirth historically precedes the governance of pregnancy and childbirth.

16 In Canada the Dominion Bureau of Statistics first used "perinatal mortality" in its analysis of figures for the year 1956 (Wadhera and Strachan 1993: 1). On England and Wales, see Armstrong 1986: 216. For an overview of 1950s national discussions related to stabilizing the category of perinatal mortality in the United States, see *History of the United States National Committee on Vital and Health Statistics, 1949–1964* (1966): 15–16.

17 Lived: this question lies outside the analytic scope of the history of the present.

18 The gendering of pastoral care is in need of genealogical study.

2 A genealogy of perinatal mortality

1 In Canada a residency programme in neonatal-perinatal medicine was first accredited by the Royal College of Physicians and Surgeons of Canada in 1987, with its initial graduates appearing in 1989 (Royal College of Physicians and Surgeons of Canada, personal communication). Some individual physicians called themselves perinatologists during the 1970s.

2 The period of the perinatal during and after birth has its own risk techniques, notably the Apgar Score (Apgar 1953). These will not be studied here as the topic is risk in pregnancy, conventionally referred to in medicine as the antenatal part of the perinatal.

3 Canadian health jurisdiction is shared between the federal and provincial levels of government, with health protection and health promotion in federal jurisdiction and health care in provincial. The health care of Aboriginal peoples and the military also falls within federal jurisdiction. The meaning of this division is constantly negotiated across levels of government.

4 The internationalism found here is limited primarily to English language sources. This is due to the entry point of the investigation, which began through an investigation of the Ontario Antenatal Record, a standardized form for antenatal risk assessment used throughout the Canadian province of Ontario from 1980 to date. Pursuing the history

of this risk text led to an international discourse about perinatal mortality reduction in medical sources of the 1950s. The chain of citations comprised a set of relations between texts – an intertextual set – encapsulated in a biomedical discourse constructed for the most part in English, secondarily in French and German.

5 Eden and Holland (1937: 631) in the *Manual of Obstetrics,* describe some of the difficulties in diagnosing fetal death during pregnancy: "The clinical diagnosis of death of the foetus *in utero* can be established only by repeated examinations. The most reliable sign is cessation of growth of the uterus, and at least a fortnight is required to determine this with certainty, the normal rate of growth being about half an inch per week. Often an actual diminution in size from absorption of the liquor amnii can be made out after some weeks. The dead foetus *in utero* feels quite different from the living; because of loss of muscle tone the limbs do not stand out distinctly and the foetus feels like a homogeneous mass. The uterus is usually lax and flaccid, and it is difficult to excite contractions by manipulating it. During the last three months, absence, on repeated examinations, of the heart-sounds is important, but no conclusion can be drawn from failure to hear them on a single occasion."

6 On the use of the infant mortality rate in late nineteenth- and early twentieth-century political argument, consult for Canada: Arnup 1994: 14–31; McLaren 1990; and Valverde 1991: 104–128; for Britain: Davin 1978; Lewis 1980; Dwork 1987; for Germany: Weindling 1988 and 1989: 241–280; for the USA: Brosco 1999 and Meckel 1990; for a French–American comparison: Klaus 1993b; for comparative studies, see Adams 1990 and McIntosh 1982 .

7 For an overview of maternal politics in the formation of social welfare states, see Bock and Thane 1991; Koven and Michel 1990; Koven and Michel 1993.

8 Brosco (1999) notes that during the late nineteenth and early twentieth centuries a "premature" baby was one lacking energy. Prematurity was not defined primarily in gestational terms.

9 John W. Ballantyne (1861–1923) received his M.D. in 1889 from the University of Edinburgh. He was appointed physician at the Royal Maternity Hospital in Glasgow (1904–1919), Professor of Obstetrics, University of Edinburgh and Lecturer on Midwifery and Gynaecology, Medical College for Women, Edinburgh. Ballantyne led the establishment of an antenatal department at the Royal Maternity Hospital, Edinburgh. His obituary in the *British Medical Journal* described his main medical contribution as "his pioneer work in seeking to establish ante-natal pathology on a scientific basis, and still more the practical application of that in the foundation of antenatal clinics for expectant mothers" (John William Ballantyne: Obituary 1923: 214).

10 See the report of the discussion on prematurity at the 1935 annual meeting of the American Academy of Pediatrics (Round Table Discussion on Prematurity 1936) in which a number of the key US researchers on neonatal mortality participated: Stewart Clifford, Ethel Dunham, Julian Hess. There was a clear consensus on prematurity as "the most important cause of infant mortality, and a very high percentage of all deaths from this cause take place in the first month" (Dunham in "Round Table Discussion on Prematurity" 1936: 109).

11 In Canada, data on stillbirths were included in provincial statistical systems effectively from their inception. A federal–provincial agreement resulted in federal reporting of stillbirths from 1921, with a stillbirth rate available from that date (Wadhera and Strachan 1993: 7). The Canadian provinces later adopted the widely used 1925 League of Nations definition of stillbirth. In England and Wales, *The Births and Deaths Registration Act, 1926*, defined stillbirth as the absence of respiration or "any other signs of life" (Sutherland 1949: 2) after birth. The criteria for stillbirth adopted in 1908, amended in 1913, by the American Public Health Association (and later the US Bureau of the Census) was consistent with the English and Welsh legal definition

(Potter and Adair 1940: 2) no evidence of life such as respiration, heartbeat and voluntary muscle movement after complete birth.

12 Cosbie (1923) was republished as a tribute in *Canadian Medical Association Journal* 1992: 1203–1207.

13 Sigismund Peller (1890–1985) was born in Tarnopol, then part of the Austro-Hungarian Empire. His M.D. was granted at the University of Vienna in 1914. After service in the Austrian army during World War I, he was resident at the Wiener Allgemeines Krankenhaus. He served as founding Director of the Bureau of Public Health of the Zionist Executive in Palestine (1926–1928). From 1929 to 1933 he worked for the Vienna Public Health Department, migrating to the United States in 1936, where he was appointed a fellow at Johns Hopkins University (1936–1940), and a research associate at New York University (1940–1943). He went into private practice in New York City in 1940.

14 Social medicine had many variants in the late nineteenth to mid-twentieth centuries (Porter and Porter 1988), with a common concern to explore the impact of social relations on the health of populations, and to advocate for a strong role for the state in improving health. Teleky's research concentrated on the occupational health of the industrial working class, studying factories and industrial hazards. He was a strong promoter of an advocacy role for medicine in securing the conditions for popular health, allying with the Austrian labour movement and demanding protective legislation. See Peller 1967: vii–viii.

15 Peller 1923a and b report on a study of 4,000 births that had taken place at the Vienna General Hospital.

16 Edith Potter M.D., Ph.D. (1901–1993). Potter took her M.D. at the University of Minnesota in 1925, followed by her Ph.D. in pathology there in 1934. She was hired as pathologist at the Chicago Lying-in Hospital in 1934. She was appointed instructor at the University of Chicago Medical School in 1934, retiring in 1967 at the rank of full professor and chief pathologist, the Chicago Department of Health. She was the first woman to be made a fellow of the American Gynaecological Society. Her book, *Pathology of the Fetus and the Newborn* (1st edition 1952), is considered a classic: "the first comprehensive presentation of perinatal pathology" (Gilbert-Barness 1995: 845). Her reputation stems mainly from her work in perinatal pathology, a field she led in establishing, as well as for her book *Rh: Its Relation to Congenital Hemolytic Disease and to Intragroup Transfusion Reaction* (Potter 1947), one of the first research publications on Rh disease.

17 Sir Dugald Baird (1899–1986) received his M.D. (Glasgow) in 1934. He was Regius Professor of Midwifery at the University of Aberdeen (1937–1966) and Obstetrician-in-Chief at the Aberdeen Maternity Hospital. From 1948 Baird began seeking out social scientists, dieticians and statisticians for his Department at Aberdeen, which became the Medical Sociology Unit after his retirement in 1965 (Jefferys 1997: 135 ftn.18). See Ann Oakley's (1984: 310–317) interview with Baird for personal reminiscence of his work. Baird also served as Vice-President of the British Eugenics Society in 1971–1974 and Director in 1977. On Baird and eugenics see http://www.eugenics_watch.com/briteugen (accessed June 2006)..

18 The subtitle of Potter and Adair's *Fetal and Neonatal Death* (1940) displays the authors' intent to demonstrate commonalities across stillbirths and neonatal deaths: *A Survey of the Incidence, Etiology, and Anatomic Manifestations of the Conditions Producing Death of the Fetus in Utero and the Infant in the Early Days of Life.*

19 Potter was initially hired in 1934 at a salary of $100 per month, doubled the following year (Hoffman 1982: 1552). Asked what was her "most important professional disappointment", Potter replied: "My only regret in Chicago, more so now than ever, was that I never had enough help. I had to carry the pathology load by myself. Any writing or researching had to be done extracurricularly, after my so-called 'working

hours', because I didn't have any time during the day. When I retired in 1967, they hired two pathologists to replace me. They assigned two or three pathology residents plus residents from the obstetrical service; they doubled the laboratory space, increased the office space, and doubled the technical and secretarial staff. The two pathologists each received a considerably higher salary than I had ... I only wish I'd asked for more money and more help, so that I could have done more research, published more about what I'd done, so that other people could use it to build on."

20 Fred L. Adair was Professor and Chair of the Department of Obstetrics and Gynecology at the University of Chicago at the time *Fetal and Neonatal Death* was published in 1940. He was head of the Chicago Lying-in Hospital (the obstetrics and gynecology teaching unit of the University of Chicago School of Medicine) from its opening in 1931.

21 From the late 1930s Potter's publications set standards for the field of fetal and neonatal pathology that was then being constituted. See Potter 1938 and 1943.

22 Papers in the 1942–1943 session of the Royal Society of Medicine ("Discussion on Stillbirth and Neonatal Mortality", 1942–1943) repeatedly addressed the recently published reports of the Toronto investigation about the effects of maternal nutrition on fetal and child health that showed stillbirth and neonatal death rates increased with poor nutrition. See Ebbs, Tisdall, Scott 1941; Ebbs *et al.* 1942.

23 Desmond (1991: 319) and Oakley (1984: 146) credit Pfaundler (1936) with introducing the term "perinatal".

24 Meinhard von Pfaundler (1876–1947) received his M.D. from the University of Graz in 1896, graduating as a pediatrician there in 1900. Between 1906 and 1939 he was Professor of Child Health at the University of Munich Faculty of Medicine, retiring as emeritus professor. He published roughly 170 articles, monographs and books, mainly in pediatrics, particularly on the diseases of children (including calcium absorption in rickets), prematurity, miscarriage and body-mass (Wiskott 1966: 101; Schulte, Michaelis and Nolte 1967: 511). He was also active in publishing on biological ideas in medicine. A biographical article (Wiskott 1966: 102) notes that von Pfaundler "did not like" the work on pediatric genetics being done at the last congress he attended in Vienna during the late 1930s, and was forced into early retirement for his lack of co-operation with fascists in Austria after the *Anschluss* (Wiskott 1966: 105).

25 For an example of Pfaundler's work on the diseases of children which was translated into English, see Pfaundler and Schlossmann 1908.

26 Biomedical research *circa* 1930–1960 had two equal and opposite tendencies with respect to birth: (1) the lightening of birth through the constitution of a perinatal interval based on commonalities between fetus and newborn, and (2) the intensifying of birth as a time of physiological transition with respect to breathing and the circulation of the blood. The work of fetal physiologists such as Barclay (1944) and Barcroft (1946) emphasized birth as physiological transition.

27 In this article, Pfaundler uses "*Frühtod*" to denote fetal demise, dating it from conception to the end of the first year.

28 The Summary Report of Meeting 4, "Foetal and Infant Mortality", the meeting at which Peller was a discussant, used the concept of "perinatal mortality" (*Proceedings of the World Population Conference, Rome, 31 August–10 September 1954* 1955: 32) as did the paper of Dr. Vasilios B. Valaoras, "Foetal, Peri-natal and Infant Mortality", *Proceedings of the World Population Conference, 1954* 1955: 323–335.

29 Peller's autobiography (1979) discusses his exposure to anti-semitism at many points in his life.

30 Titmuss and Baird were personal friends and research collaborators (Oakley 1984: 310–311).

31 Yankauer's (1950 and 1953) previous studies were posed in terms of "fetal and infant mortality". Nancy Krieger (2001) notes that Yankauer (1950) was the first to use the

term "social epidemiology". He received his M.D. from Harvard Medical School (1938), later teaching at Harvard in the School of Public Health (see "Yankauer, Alfred" in Cattell 1975: 1124). Yankauer was appointed Professor in Community Medicine at the University of Massachusetts (1973) and served as an editor of the prestigious *American Journal of Public Health.*

32 Dugald Baird reanalysed the data of the National Birth Trust Study, *Perinatal Mortality* (1963), concluding that the data supported hospital births as safer than home births and also supported induction at forty-one weeks (Tew 1990: 245–248). See Tew's critique (ibid.) of Baird's analysis.

33 Standardized antenatal risk assessment forms are currently used in all Canadian provinces and territories (Hall 1994: 1240, Table 1).

34 The Ontario Antenatal Record has been subject to ongoing major and minor revisions since 1980, the last major revision occurring in 2000 (Chance 1992; OMA Subcommittee on the Antenatal Record 2000).

35 *Perinatal Problems* (Ontario Council of Health 1971) noted that the Canadian provinces of Alberta and Nova Scotia already had pregnancy risk screening in place.

36 The main work in devising the Antenatal Record had been done by the perinatologist Dr Graham Chance, who had obtained funding from the Foundation of the Hospital for Sick Children (Toronto) to develop and field test it, then obtaining its endorsement by the Ontario Medical Association Special Committee on Perinatal Care, of which he was a member.

37 In North America, Nesbitt and Aubrey (1969) is likely the best known of the risk scores that were devised during the late 1960s and found throughout the 1970s. For an overview of risk-scoring devices see Alexander and Keirse 1989.

38 The first widely accepted international studies showing the benefits of regional perinatal care were produced in Quebec (Comité d'étude de la mortalité périnatale 1967 and 1969).

39 Three years prior to the 1979 Ontario report, *A Regionalized System for Reproductive Medical Care in Ontario*, the American Medical Association had approved its own proposal for the regionalization of perinatal care, *Towards Improving the Outcome of Pregnancy*, with widespread support among physicians (Perkins 1993).

40 A survey done in 1989–1990 of all Ontario hospitals offering maternal and newborn care estimated that 3,500 of the total 132,172 provincial births that year received no prenatal care, and thus had no documentation for the pregnancy prior to birth (Hanvey *et al.* 1991: 45, Table 2–11). Antenatal risk governance is effective at the level of conduct.

41 Modified from Last 2001: 159 (entry on "risk"). Last's *Dictionary of Epidemiology* was edited for the International Epidemiology Association and is a standard reference source.

42 Modifiable and non-modifiable risk determinants are also often contrasted with "risk markers", which are attributes or exposures correlated with an outcome.

43 The Conference, "Controlling the Risk: Science to Combat Global Infectious Diseases," was held in Toronto, 9–19 November 2004, sponsored by the Centre for Global Health Research (affiliated with St Michael's Hospital).

44 The point about iatrogensis and unnecessary interventions is often argued through the example of glucose testing. Routine glucose tolerance testing for gestational diabetes produces results that are not reproducible in 50–70 per cent of cases, and the standard of care, insulin, has not been shown to improve neonatal outcomes (Hunter and Keirse 1989: 419–439).

45 The representations of the perinatal interval found in the "Perinatal Period Chart" (1959) produced by the Committee on Maternal and Child Care, American Medical Association (Figure 2.3) and Pfaundler's 1936 graph of perinatal mortality (Figure 2.2) bear many resemblances.

3 Health beyond risk: a midwifery ethos in prenatal care

1 Physiological, healthy and normal are of course by no means equivalent terms, but their differentiation lies beyond the scope of the argument here. On the distinction between the "abnormal" and the "pathological", see Note 26 below.
2 For a comprehensive and reliable overview of Canadian midwifery history during the 1970s and 1980s, see Bourgeault, Benoit and Davis-Floyd 2004b.
3 Prior to the legal recognition of midwifery as a health care profession, individual midwives were potentially both personally and criminally liable for harm to clients occurring during pregnancy or the birth process.
4 *Midwifery Act, 1991*, S.O. 1991, C. 31, hereinafter *Midwifery Act*. The passage of this legislation occurred in the context of a provincial review of legislation governing the health professions and the passage of a new *Regulated Health Professions Act, 1991*, S.O. 1991, C. 18.
5 "Continuity of care" denotes the provision of care during birth by health care providers previously known to the woman in labour. This is not assured in medical practice models that group eight to twelve physicians in an on-call group. The usual practice model among midwives is to have two caregivers follow a woman in prenatal care, although this occasionally rises to four.
6 The Society of Obstetricians and Gynecologists of Canada (1997) condemns home birth as unsafe. Choice of birthplace has been fundamental to community midwifery in Canada and vigorously opposed by the professional organizations of physicians.
7 As of 1998, all Ontario hospitals required the submission of the Antenatal Record for hospital births.
8 A minority of midwives had worked with the Antenatal Record prior to 1994: (1) those associated with the midwifery project at McMaster-Chedoke Hospital (Harvey, Kaufman and Rice 1995) where the form had been used since the mid-1980s and (2) a number of midwives whose clients entered hospital under the admitting privileges of physicians wishing to abide by hospital protocols.
9 Across Canada in 1989, 95.4 per cent of women during pregnancy received prenatal care from a registered health care provider, with an additional 0.7 per cent seeing a traditional birth attendant (Buekens 1995: 192).
10 The 1992 revision of the Antenatal Record was in use during the period of research. It continued to be employed until March 2002 when it was gradually replaced as caregivers ordered new stock from the Ministry of Health. On the revision of 2000 see OMA Subcommittee on the Antenatal Record 2000.
11 In a regionalized system of health care, the most basic levels of care are located close to clients, while more complex levels are offered in the central area of each region.
12 Midwifery research participants were geographically distributed throughout Ontario, with eight from the Greater Toronto Area (where the majority of midwives practised), five from northern Ontario, three from Western Ontario, and four from Eastern Ontario. These were chosen from the 1996 list of registered midwives provided by the College of Midwives of Ontario, with all 1995 participants being members of the 1996 list. Sixty-eight midwives were registered in 1996, with fifty-eight in active practice. All were women, as the sole man who was a registered midwife had left practice in the spring of 1996. I conducted all midwifery interviews, fourteen in person and six by telephone. The interviews lasted from thirty to sixty minutes and were tape recorded and later transcribed.
13 Physicians were recruited through contacts known to me: colleagues, midwives, physicians and a physiotherapist. An additional three physicians were recruited by approaching a hospital research committee. The physician comparison group consisted of ten men and ten women, seven from the Greater Toronto Area, four from Northern Ontario, four from Western Ontario, and three from Eastern Ontario. Three worked in

rural areas. Four worked in community health centres. Eleven had admitting privileges at a secondary care facility and nine at a primary care facility.

14 The Transitional Council of the College of Midwives of Ontario existed from 14 February 1993 to 31 December 1993. During that time it devised standards and organizational infrastructure in preparation for the proclamation of the *Midwifery Act, 1991* (Bourgeault 1996: 87).

15 Although hospital protocols require completion of the Antenatal Record, midwives are legally and professionally responsible for their records content to the "Standard on Records Content" of their College (College of Midwives of Ontario 1994b). Through this practice standard the College prescribes the minimum of what is to be recorded, but does not attempt to make a single set of records compulsory nor to prescribe content exhaustively. The Standard on Records Content does not require that a dietary history, familial obstetrical history, prior labour history or more detailed information on menstrual history be taken. College practice standards are less stringent than community ones.

16 "Charting" refers to keeping individual records on a patient/client in care.

17 During the period of the interviews, 1995–1998, Ontario midwives were engaged in ongoing internal discussions about risk. Four of the research participants remarked that concerns about the appropriateness of using risk in relation to midwifery care were being raised among Ontario midwives. Three participants sent me hard-to-find documents related to midwifery and risk assessment (Chatworth 1996; Oakley and Houd 1990; Saxell 1994).

18 The risk factors that the physicians suggested as not appropriate for referral or consultation included history of prior stillbirth, neonatal death, primagravida (women giving birth for the first time over 35 years), grand multipara (a woman with multiple births), non-prescription drugs, anaemia not responding to treatment, and cigarette smoking.

19 One physician suggested that Level B be divided in two: increased risk with and without referral. Five of the physicians viewed Level B risk factors on a scale of seriousness between Levels A and B.

20 "Policy Statement on Records", Transitional Council of the College of Midwives, approved May 1993. Records of the College of Midwives of Ontario.

21 Many francophone hospitals are called the *Hôtel Dieu*. It is a generic name in this quotation rather than referring to a specific hospital.

22 The provisional 2000 edition of the Antenatal Record was modified to formally include midwives.

23 In only two places on Antenatal Record 1 and 2 is there possibility for a kind of exchange outside the questioner–respondent format. This occurs in the "Discussion Topics" in the lower right-hand corners of Antenatal 1 and 2, which, given the semiotic conventions of textual space, signify this area as the least important aspect of each visit. "Discussion Topics" mark an educational component within the Antenatal Record.

24 *Midwifery Act*, *supra* n. 4, s. 3.

25 *Medicine Act, 1991, S.O. 1991*, c. 30; hereinafter *Medicine Act*.

26 "Abnormal" rather than "pathological" appears in the *Medicine Act, 1991*, Section 4(1). The "abnormal" and the "pathological" are not conceptually equivalent. A state of health may include quantitative laboratory results that are "abnormal" in the sense of measuring above or below the mean/reference range. The reverse is also the case: laboratory tests falling within the reference range ("normal") may be compatible with pathology. Thus "abnormal" may indicate a state of health, not pathology. On the distinction between the abnormal and the pathological see Canguilhem 1989: 151–180. I will continue to use "abnormal" as this is the legislated terminology.

27 The 1992 revision was produced by the Reproductive Care Committee of the Ontario Medical Association (Chance 1992: 23).

28 Unlike the majority of the midwives at the time of interview, eight physicians were subject to hospital chart audits that checked their compliance in completing the Antenatal Record; an additional four physicians either had participated or were currently participating in the audits. Another four physicians were not subject to chart audits, but had experienced them during residency or during the initial period after obtaining hospital privileges. Ten of the physicians mentioned in-hospital guidelines for consultations and referrals that take precedence over the Antenatal Record in cases of conflict. Institutional controls on physicians governing completion of the Antenatal Record thus exceeded those on midwives.

29 *Taber's* is the standard dictionary used by health professionals other than physicians, who use *Dorland*. There is no MESH listing (Medical Subject Heading, National Library of Medicine, Bethesda, Maryland) for "indication".

30 The Association of Ontario Midwives was a membership organization of practising midwives formed during the 1980s.

31 Records of the College of Midwives of Ontario, Committee on Standards and Professional Relations, 1993.

32 Attachment A, "Guidelines to Scope of Practice", Minutes of Nov. 8, 1990, Meeting of Standards and Qualifications Committee, Interim Regulatory Council on Midwifery, Records of the College of Midwives of Ontario.

33 "Indications for Mandatory Discussion, Consultation and Transfer of Care", draft September 15, 1993, Transitional Council of the College of Midwives, Records of the College of Midwives of Ontario.

34 The College of Physicians and Surgeons of Ontario and the Ontario Medical Association were consulted during the drafting of the "Indications for Mandatory Discussion, Consultation and Transfer of Care". Because the "Indications" set the standard for consultation and transfer of care between midwives and physicians, professional co-operation was called for both legally and organizationally.

35 François Ewald has emphasized the intrinsic relation between norms and standardization, with the norm being "a measurement and a means of producing a common standard" (Ewald 1990: 141). Normalization is a mechanism for producing societies characterized by standardization.

4 Legal fiction and reality effects: evidence of perinatal risk

1 "Child *en ventre sa mère*" is a specialized legal term in English, a rendering that omits the "de" that would be expected in French: the "child *en ventre de sa mère*". *Ventre* may be translated as belly or womb depending on context of usage.

2 "Civil law" is ambiguous in Anglo-Canadian usage. It is used to contrast French Canadian, codified law with English Canadian common law. Civil law is also distinguished from criminal law.

3 Tort law deals with civil liability for losses/harms that have happened as a result of a breach in non-contractual duty. An example would be suing someone for damages as the result of being impaled by an icicle falling from the roof of a building. The hazards of thresholds are multifarious.

4 Jonathan Simon (1987, 1988) led the field in his studies of insurance and "actuarial practices". His earlier work (including his co-publications with Malcolm Feeley [Feeley and Simon 1992, 1994]) has examined the impact of insurance as a form of risk/actuarial reasoning on criminal law and punishment regimes. Simon's enormously influential work is clearly at play in my comments on "actuarial effects", although my concern lies outside criminal law.

5 *Montreal Tramways v. Léveillé,* [1933] S.C.R. 456, 41 C.R.C. 291, 4 D.L.R. 337. Hereinafter *Montreal Tramways* cited to D.L.R.

6 The right of children born posthumously to inherit real property has been recognized for centuries in English common law, vide *The Earl of Bedford's Case*, [1587] 7 Co. Rep. 7b, 77 E.R. 421 (Court of Wards) (hereinafter *Earl of Bedford*). Commentators on French and Quebec civil law customarily link *l'enfant conçu mais non encore né* (the child conceived but not yet born) to Ancient Roman law of inheritance. For discussion see *Montreal Tramways*, *supra* n. 5 at 341–343.

7 A comprehensive analysis of the provisions in the *Civil Code of Lower Canada* dealing with the patrimonial interests of the unborn child and the appointment of a curator to protect its economic interests is found in *Tremblay v. Daigle*, [1989] 2 S.C.R. 530 at 556–565. Hereinafter *Daigle.*

8 Means (1968: 422) makes repeated reference to "the fetus" and treats pregnancy as having a standard length of forty weeks in the late medical and early modern periods. These aspects of his argument are anachronistic as until the late eighteenth century, pregnancy was not regarded as having a standard length. So too, the concept of the biological fetus dates to Soemmerring in 1799.

9 In a particularly clear and careful passage, *Elliot v. Lord Joicey,* [1935] A.C. 209 at 210 (H.L.(E.)) summarized *Villar v. Gilbey*, [1907] A.C. 139 (H. L.(E.)) on the distinction between the "ordinary" vs. the "fictional" meaning of words with respect to the child *en ventre sa mère*: "1. Words referring to children or issue 'born' before or 'living' at, or 'surviving', a particular point of time or event will not in their ordinary or natural meaning include a child en ventre sa mère at the relevant date. 2. The ordinary or natural meaning of the words may be departed from, and a fictional construction applied to them so as to include therein a child en ventre sa mère at the relevant date and subsequently born alive if, but only if, that fictional construction will secure to the child a benefit to which it would have been entitled if it had been actually born at the relevant date. 3. The only reason and the only justification for applying such a fictional construction is that where a person makes a gift to a class of children or issue described as 'born' before or 'living' at or 'surviving' a particular point of time or event, a child en ventre sa mère must necessarily be within the reason and motive of the gift."

10 On awe in English common law, see Goodrich 1990: 231.

11 *Dig.* 1.5.7 (eds. Mommsen and Krueger).

12 *Dig.* 1.6.26 (eds. Mommsen and Krueger).

13 The *Digest of Justinian* on *qui in utero est* is quoted in *Walker v. Great Northern Railway Co. of Ireland*, (1891), 28 L.R.Ir.69 at 83 (Irish Court of Queen's Bench) (hereinafter *Walker*) and *Montreal Tramways, supra* n. 5 at 341.

14 *Earl of Bedford*, *supra* n. 6.

15 *Thelusson v. Woodford,* (1815), 1 Ves. Jun. 112, 32 E.R. 1030 at 1040–1042 (H. L. (E)).

16 *Borowski v. Canada (Attorney General)* (1987), 39 D.L.R. (4th) 731 at 743 (Sask. C.A.); *Daigle*, *supra* n. 7 at 569.

17 *Re Charlton Estate*, [1919] 1 W.W.R. 134 (Man. K.B.).

18 *Re Sloan Estate*, [1937] 3 W.W.R. 455 (B.C.S.C.).

19 I have sketched the place of the unborn child fiction in the 1866 *Civil Code* rather than the 1994 *Civil Code* because the case law examined in this chapter was tried prior to the passage of the 1994 revision. The provisions in the *Civil Code of Quebec* (1994) dealing with *l'enfant conçu mais non encore né* revised the *Civil Code of Lower Canada* (1866) in the light of the Supreme Court of Canada judgement in *Daigle* (*supra* n. 7) to clarify that the fiction of the unborn child conferred patrimonial interests, not extrapatrimonial ones such as a "right to life". The regime of tutorship and guardianship of the unborn child was also revised in the 1994 *Civil Code* consonant with the Supreme Court of Canada decision in *Daigle* that a "fetus" was not

recognized as a juridical person in the *Civil Code of Lower Canada* and the *Quebec Charter of Human Rights and Freedoms.*

20 Art. 838 C.C.L.C.

21 Art. 345 C.C.L.C.

22 Art. 608 C.C.L.C.

23 Art. 771 C.C.L.C. Arts. 948 and 2543 C.C.L.C. also dealt with the patrimonial interests of the unborn child.

24 Doran (1996) argues that the concept of "accident" represents a device used by factory inspectors for constituting workers' health status and injuries on the job as reportable phenomena. "Accident" was thus an official notion, not the perspective of workers for whom injury and illness were no accident but rather a systemic property of the conditions of their labour.

25 *Giddings v. Canadian Railway* (1920), 53 D.L.R. 3 (Sask. C.A.). Hereinafter *Giddings.*

26 *Act Respecting Compensation to the Families of Persons Killed in Accidents,* R.S.S., 1909, 135.

27 *Giddings*, *supra* n. 25 at 9.

28 *Chapman v. C.N.R. and Parry Sound*, (1943) 2 D.L.R. 98 (Ont. H.C.).

29 *Workmen's Compensation Act,* R.S.O. 1937, c. 204.

30 *Fatal Accidents Act,* R.S.O. 1937, 5.8(3).

31 *Fitzsimonds v. Royal Insurance Company of Canada*, (1984) 7 D.L.R. (4th) 406 (Alta. C.A.).

32 Ibid. at 409 and 412.

33 The "unborn child" has been used in Canadian family law to award custody during pregnancy, again contingent on live birth. The decisions have not been risk managerial in their discursive form. See *K. v. K.*, (1933) 3 W.W.R. 351 (Man. K.B.); *Solowan v. Solowan*, (1953) 8 W.W.R. 288 (Alta. S.C.).

34 *Montreal Tramways*, *supra* n. 5.

35 The leading precedent in the USA was *Bonbrest v. Kotz*, 65 F. Supp. 138 (D.D.C. 1946) (hereinafter *Bonbrest)* and, in Australia, *Watt v. Rama,* (1972) V.R. 353 (Vic. S.C.) (hereinafter *Watt*). For statutory provisions in the United Kingdom, see n. 54 (below).

36 *Dietrich v. Northhampton*, 52 Am. R. 242 (1884). Hereinafter *Dietrich.*

37 *Walker*, *supra* n. 13.

38 *Dietrich*, *supra* n. 36 at 242–243.

39 Ibid.

40 Ibid. at 244.

41 For commentary on *Dietrich* see Santello 1988: 749–750.

42 *Dietrich*, *supra* n. 36 at 245.

43 Ibid.

44 *Walker, supra* n. 13 at 71.

45 Ibid. at 71.

46 Ibid. at 70.

47 Ibid. at 83.

48 Ibid. at 83.

49 Ibid. at 81. "Lusus naturae" means sport of nature, and "lusus scientiae" sport of science. The passage is simultaneously bombastic and prescient.

50 *Montreal Tramways*, *supra* n. 5.

51 *Bonbrest*, *supra* n. 35.

52 *Watt*, *supra* n. 35.

53 *Duval v. Séguin* (1972), 26 D.L.R. (3d) 418 (Ont. H.C.), aff'g (1973), 40 D.L.R. (3d) 666 (Ont. C.A.). Hereinafter *Duval.*

54 In the UK, the *Congenital Disabilities (Civil Liability) Act*, *1976*, was passed in the wake of the thalidomide disaster. It established the capacity of individuals to sue for prenatal injuries (Conaghan and Mansell 1993: 89).

55 Art, 1053 C.C.L.C. cited in *Montreal Tramways, supra* n. 5 at 339
56 Ibid. at 346.
57 Ibid. at 344.
58 *Donoghue v. Stephenson*, [1932] All E.R. Rep. 1 (H.L.(E)). Hereinafter *Donoghue*.
59 *Montreal Tramways, supra* n. 5 at 356.
60 Ibid. at 353.
61 Ibid. at 353.
62 Ibid. at 345.
63 Ibid. at 366.
64 Ibid. at 345.
65 Ibid. at 352.
66 *Farmer v. Canada Packers Ltd.*, [1956] O.R. 657, 6 D.L.R. (2d) 63 (Ont. S.C.).
67 *Montreal Tramways, supra* n. 5 was discussed at length in *Watt, supra* n. 35 at 353 and cited in *Duval, supra* n. 53, at 430–431 (Ont. H.C.) The frequent application of *Montreal Tramways* in common law has been noted in *Dobson (Guardian Ad Litem) v. Dobson* , [1997], N.B.J. No. 232 at para. 5 (N.B.C.A.) (Q.L.).
68 Although Priest (1990) dates the reconfiguration of tort law as a regime of risk management to the 1960s, aspects of it were formed during the 1930s with the creation of "reasonable foreseeability" as the standard for the duty of care.
69 *Donoghue, supra* n. 58.
70 Subsequent precedents extending the reach of proximity and cited in *Duval, supra* n. 53 were: *Grant v. Australian Knitting Mills, Ltd.,* [1936] A.C. 85 P.C.; *Dorset Yacht Co. v. Home Office*, [1970] A.C. 1004 (H.L.(E)).
71 *Duval, supra* n. 53 at 431–432 (Ont. H.C.) citing *Donahue, supra* n. 58 at 11.
72 (Ont. H.C.)
73 Ibid. at 432.
74 Ibid. at 434.
75 *Dehler v. Ottawa Civic Hospital* (1979), 25 O.R. (2d) 748, 101 D.L.R. (3d) 686 (Ont. H.C.), aff'd 117 D.L.R. (3D) 512 (Ont. C.A.)
76 *Seede v. Camco Inc.* (1985), 50 O.R. (2d) 218 (Ont. S.C.), aff'd (1986), SS O.R. (2d) 352, app. for leave to appeal dismissed by the Supreme Court of Canada. Hereinafter *Seede* cited to O.R.
77 *Winnipeg Child and Family Services (Northwest Area) v. G.(D.F.)*, [1997] 3 S.C.R. 925.
78 *Seede, supra* n. 76 at 221 (Ont. G.C.).
79 Wrongful birth actions are brought by the parents or guardians of a disabled child; wrongful life when the child her/himself is the plaintiff. In wrongful birth, the claim is that the child would not have been born but for the physician's negligence. In wrongful life suits, the child seeks damages for pain and suffering as well as lifelong heath care costs. In Britain, wrongful life suits were banned by the *Congenital Disabilities Act, 1976*, and the majority of courts in the United States have denied such claims. In Canada, see *Arndt v. Smith*, [1995] B.C.J. No. 1416 (B.C.C.A.) (QL) and *Cherry (Guardian ad Litem of) v. Borsman*, [1992] B.C.J. No. 1687 (B.C.C.A.) (QL).
80 *Duval, supra* n. 53 at 433 (Ont. H.C.).
81 *Bonbrest, supra* n. 35; *Watt, supra* n. 35.
82 *Duval, supra* n. 53 at 434 (Ont. H.C.).
83 Ibid. at 425.
84 Ibid. at 426.
85 Ibid. at 426.
86 *Duval, supra* n. 53 at 669 (Ont. C.A.). In this passage from the appeal of *Duval* supra n. 53, Justice Osler quotes the testimony of the neurologist, Dr E.A. Atack, from the transcripts of the original trial.
87 *Duval, supra* n. 53 at 434 (Ont. H.C.).

88 Ibid. at 433.
89 Ibid.
90 These categories of analysis are from systemic functional grammar. See Halliday 1985.
91 *Dobson (Litigation Guardian of) v. Dobson,* [1999] 2 S.C.R. 753 at 788–789.
92 The conflation of the unborn child fiction with the biological concept of the fetus was also found in the judgements of the Superior Court and the Quebec Court of Appeal in the Daigle case (*Tremblay v. Daigle*, [1989] R.J.Q. 1980 (S.C.); *Tremblay v. Daigle*, [1989] 59 D.L.R. (4th) 609 (C.A.)). I have not examined the judgements in *Daigle* because neither the evidence nor the legal reasoning in the decisions was argued in terms of *risk* to a child's health from a prenatal injury. In *Daigle*, a man who had helped conceive a child sought an injunction to prevent his ex-partner from having an abortion. The action was brought under provincial human rights legislation (*Quebec Charter of Rights and Freedoms*), Quebec civil law, and the *Canadian Charter of Rights and Freedoms*.

Daigle is pertinent here because the articles in the *Civil Code of Lower Canada* dealing with the guardianship of the unborn child together with its patrimonial interests were argued by counsel for Tremblay to interpret the "foetus" as a juridical person. The trial judge and the majority decision in the Quebec Court of Appeal equated the unborn child fiction of the civil law with the biological concept of the fetus: a deeply ahistorical reading. When the temporal fiction of the unborn child is fused with the biological concept of the fetus, claims made about the fetus such as human status and, by inference, legal personhood, apply also to the unborn child, thereby displacing centuries of legal reasoning. The Supreme Court of Canada decision in the *Daigle* appeal (*Daigle, supra* n. 7) argued that the *Civil Code* provisions about curatorship and patrimonial interests existed solely to protect the economic interests of the future child, and thus did not confer juridical personality on the "foetus" – an "extrapatrimonial interest". Patrimonial rights were realized only when the "fetus" was born alive and viable: "If the foetus is not born alive and viable then the interests referred to in these articles [of the *Civil Code*] disappear, as if the foetus did not exist at all" (ibid. at 560). Elsewhere in the same judgement, the Supreme Court strongly asserted the distinction between legal and biological reasoning about the "foetus" (ibid. at 553).

On the juridical status of "*l'enfant non encore né*" in the *Civil Code of Lower Canada* see Rivet 1978. For discussion of the legal decisions in *Daigle*, see Greschner 1990; Rhéaume 1990; Schaffer 1994.
93 For the configuration of the legal subject with respect to uncertainty in contract law, see O'Malley 203: 236 and 2004.

5 Child welfare at the perinatal threshold: making orders protecting fetuses

1 *Borowski v. Canada (Attorney General)* (1987), 39 D.L.R. (4th) 731, [1987] 4 W.W.R. 385 (Sask. C.A.), aff'g (1983), 4 D.L.R. (4th) 112, [1984] 1 W.W.R. 15 (Sask. Q.B.). Hereinafter *Borowski*. The Supreme Court of Canada granted leave to appeal but exercised its discretion not to decide the appeal on the grounds that it was moot: *Borowski v. Canada (Attorney General)*, [1989], 1 S.C.R. 342.
2 *Tremblay v. Daigle,* [1989] 2 S.C.R. 530. Hereinafter *Daigle*.
3 Within the criminal law, the key cases involved the line of demarcation between the fetus and the subject forms of "person" and "human being". In *R. v. Sullivan and Lemay* (1991), 63 C.C.C. (3d) 97 (S.C.C.), aff'g (1988), 31 B.C.L.R. (2d) 145, 65 C.R. (3d) 256 (B.C.C.A.), rev'g (1986), 55 C.R. (3d) 48, 31 C.C.C. (3d) 62 (B.C.S.C.) (hereinafter *Sullivan and Lemay*), two midwives were charged under the *Criminal*

Code with criminal negligence causing death of a person after a labour they were assisting resulted in a perinatal death. The trial judge concluded that the fetus was not a "human being" under s. 206(1) of the *Criminal Code*, but that it was a "person" under s. 203, and convicted the midwives of criminal negligence causing death of a person. The decision was reversed in the subsequent appeals to the British Columbia Court of Appeal and the Supreme Court of Canada, where it was decided that human being and person were used synonymously in the *Criminal Code*, and that the fetus only becomes a "person" and a "human being" in a legal sense when it "has completely proceeded, in a living state, from the body of its mother". (*Sullivan and Lemay* (1991), 63 C.C.C. (3D) 97 at 106 (S.C.C.) In the case of *R. v. Drummond*, [1996] O.J. No. 4597 (Ont. Ct. Justice (Prov. Div.)) (QL), a woman was charged with attempted murder after shooting herself through her vagina with a gun when thirty-six weeks pregnant. The child was born alive and underwent surgery for the removal of pellets from his brain. The case was dismissed on the grounds that, prior to live birth, the fetus was not a person within the meaning of the *Criminal Code*.

4 In Canada, "Aboriginal" denotes three differing groups: First Nations (status and non-status Indian), Inuit and Métis. Métis originally has the sense of joint Settler and non-Settler parentage.

5 Those accused of child neglect have primarily been mothers rather than fathers (Swift 1995: 10).

6 Jurisdiction in Aboriginal child welfare is split between provincial and federal authorities. Aboriginal reserves fall under federal jurisdiction, but child welfare is a provincial matter. During the 1980s and continuing to date, a separate regime of Aboriginal child and family services was created (Gray-Withers 1997; Tait 2000) in an attempt to bring Aboriginal child welfare under Aboriginal control.

7 *Child Welfare Act*, R.S.O. 1980, c. 66, s. 19.

8 *Child and Family Services Act, 1984,* S.O. 1984, Pt. 3 c.55, s. 37, 2 (b), (d), (g).

9 *Fam. & Children's Services of London and Middlesex v. G. (D.)* (1989), 20 R.F.L. (3rd) 429 (Ont. Fam. Ct.). Many thanks to Roxanne Mykitiuk for alerting me to this case.

10 *Joe v. Y.T. Dir. of Fam. and Children's Services* (1986), 5 B.C.L.R. (2d) 267 (Y.T.S.C.). Hereinafter *Joe*. The trial decision was not reported, but the decision was partially reproduced in the appeal decision. For an overview of how "fetal alcohol syndrome" has been deployed with respect to Aboriginal people in the USA, see Dineen 1994.

11 *Children's Act,* S.Y.T. 1984 , c.2, Pt. 4, SS. 134(1).

12 *Ibid.*

13 The New Brunswick *Child and Family Services and Family Relations Act*, S.N.B. 1980, c. C-2.1(g)(a)(b) encompassed "unborn child" and "stillborn child" with the meaning of "child", but contained no statutory language pertaining to risk. One case was reported under this legislation, *Nouveau-Brunswick (Ministre de la Santé et des Services communautaires) v. A.D.* (1990), 109 N.B.R. (2d) 192, 273 A.P.D. 192 (N.B.Q.B (Fam. Div.). A six-month supervisory order over a woman and her "unborn child" was granted to child welfare authorities, with requirements that the woman receive medical care and comply with a hospital birth.

14 Deputy Territorial Court Judge quoted in *Joe, supra* n. 10 at 270–271.

15 *Joe, supra* n. 10 at 271.

16 Ibid. at 269.

17 *Re Children's Aid Society for the District of Kenora and J.L.* (1981), 134 D.L.R. (3d) 249 (Ont. Prov. Ct. (Fam. Div.)). Hereinafter *Kenora*.

18 Ibid. at 251.

19 Ibid. at 252.

20 Ibid.

21 *Child Welfare Act*, R.S.O. 1980, c. 66, ss. 19(1)(b)(ix) and (xi).

22 *Kenora, supra* n. 17 at 252.

23 "Fetal apprehension" extends "child apprehension", that is, the power of child welfare authorities to detain children. "Fetal apprehension" does not carry the sense of fetal anxiety or concern. As with "antenatal record", it is used utterly without irony.
24 In indirect discourse, the speaker attributes a statement to another speaker without directly quoting her/him, as in "The doctor said she had fetal alcohol syndrome." Indirect discourse enables the introduction of the speech/texts of others into a statement, thus incorporating and potentially transforming that knowledge. The transformations of the original speaker's/author's statement knowledge may occur in ways not recoverable by the addressee of the indirect discourse. In the case law examined here, judicial decisions incorporate medical evidence through indirect discourse.
25 Many thanks to Lachlan Story for his generosity in briefing me on the research history of FAS/FAE.
26 *British Columbia (Superintendent of Family and Child Services) v. McDonald* (1982), 135 D.L.R. (3d) 330 (B.C.S.C.). Hereinafter *McDonald.*
27 Ibid. at 331.
28 Ibid. at 332.
29 *Family and Child Service Act*, R.S.B.C. 1980, c. 11, s. 1.
30 *McDonald*, *supra* n. 26 at 333.
31 Ibid. at 335.
32 *Re Children's Aid Society of Belleville, Hastings County and T. et al.* (1987), 59 O.R. (2d) 204. (Ont. Prov. Ct. (Fam. Div.)). Hereinafter *Belleville.*
33 *Re R.* (1987), 9 R.F.L. (3d) 415 (B.C.Prov. Ct. (Fam. Div)). Hereinafter *Re R. Re Baby R* (1988), 15 R.F.L. (3d) 225 (B.C.S.C.). Hereinafter *Re Baby R.*
34 *Belleville*, *supra* n. 32 at 204–205.
35 Ibid. at 205.
36 Ibid. at 205–206.
37 Ibid. at 206.
38 Ibid. at 204.
39 Ibid.
40 Ibid.
41 *Child and Family Services Act, 1984,* S.O. 1984, c. 55, s.37(2)(b) (e).
42 *Belleville, supra* n. 32 at 205.
43 Ibid. at 206.
44 *Re R*, *supra* n. 33 at 416.
45 *Re Baby R*, *supra* n. 33 at 230.
46 Footling breech presentation was and is regarded as an intrapartum (during labour) risk factor, an indication for routine Caesarian section although contested by midwives internationally. See Hannah, Hannah, Newson, Hodnett, Saigal and Willan (2000): Hofmeyer and Hannah (2006).
47 *Re R*, *supra* n. 33 at 420.
48 Ibid.
49 Ibid. at 417.
50 Ibid. at 419.
51 *Re Baby R*, *supra* n. 33 at 234.
52 Ibid. at 229.
53 Ibid. at 234–235.
54 *Sullivan and Lemay*, *supra* n. 3.
55 *Re F (in utero)*, [1988] 2 All E.R. 193 (C.A.)
56 *Re Baby R*, *supra* n. 33 at 235–236.
57 *Re R*, *supra* n. 33 at 416.
58 *Borowski*, *supra* n. 1; *R. v. Morgentaler*, [1988] 1 S.C.R. 30.

59 *Re Baby R*, *supra* n. 33 at 231. *Baby R* attracted more academic and legal interest than had previous fetal apprehension cases (see Dawson 1990; Diduck 1993; Maier 1989). The decision of the appellate court appeared with two case comments, D. Majury (1987) 15 R.F.L. (3d) at 225 and D.W. Phillips (1987) 15 R.F.L. (3d) at 228.
60 *Re Baby R*, *supra* n. 33 at 231.
61 Ibid. at 237.
62 Ibid.
63 *Re A (in utero)* (1990), 75 O.R. (2d) 82 (Ont. U.F.C.).
64 Ibid. at 86. Level II and Level III indicate secondary and tertiary care hospitals that are used for births judged high risk and very high risk.
65 Ibid. at 89.
66 Ibid. at 91–92.
67 For discussion, see *Report of the Royal Commission on Aboriginal Peoples* (1996), Vol. 3, *passim*.
68 *Child and Family Services of Winnipeg Northwest v. D.F.G.* (1996), 111 Man. R. (2d) 219 (Man. Q.B.); *Winnipeg Child and Family Services (Northwest Area) v. D.F.G.*, (1996) 113 Man. R. (2d) 3 (Man. C.A.); *Winnipeg Child and Family Services (Northwest Area) v. G.(D.F.)*, [1997] 3 S.C.R. 925. Hereinafter *Winnipeg*.
69 *Winnipeg, supra* n. 68 (S.C.C.) (*Factum of the Appellant*, Winnipeg Child and Family Services, Northwest Area, to the Supreme Court of Canada at para. 27).

In the Canadian legal regime, a factum is a document containing the written legal argument of a party or intervenor in a case. As arguments from facta often appear in legal judgements, the submission of a factum is an opportunity for a party or an intervenor to persuade the court of a particular position. Courts have the sole discretion to grant intervenor status to those who apply. An intervention can be allowed at any level of the case. For discussion and examples of facta filed by the main feminist advocacy organization in Canadian constitutional law, see Women's Legal, Education and Action Fund 1996.

70 *Winnipeg*, *supra* n. 68 (S.C.C.) (*Factum of the Intervenors* Women's Health Clinic Inc., Métis Women of Manitoba Inc., Native Women's Transition Centre Inc., Manitoba Association of Rights and Liberties to the Supreme Court of Canada at paras 12–15.This Factum is available at http://women'shealthclinic. org/resources/pwatnto/9–Brief.html (accessed May 2006).
71 Ibid. at para. 16.
72 *Winnipeg*, *supra* n. 68 at 933 (S.C.C.).
73 *Winnipeg*, *supra* n. 68 at 6 (Man. C.A.).
74 Ibid. at 6.
75 Ibid. at 7.
76 *Winnipeg*, *supra* n. 67 at 926 (S.C.C.).
77 *Winnipeg*, *supra*. n. 68 (S.C.C.) (*Factum of the Appellant* at para. 2).
78 Ibid. at para. 77.
79 Ibid. at para. 35.
80 Ibid. at para. 36.
81 Ibid. at para. 40.
82 Ibid. at para. 43.
83 Ibid. at para. 48.
84 *Winnipeg*, *supra* n.68 (S.C.C.) (*Factum of the Intervenor*Women's Legal Education and Action Fund to the Supreme Court of Canada at para. 31).
85 *Winnipeg*, *supra* n. 68 at 941 (S.C.C.).
86 Ibid. at 954.
87 Ibid. at 943.
88 Ibid.
89 Ibid. at 945.

90 Ibid. at 981.
91 Ibid. at 986. The dissent in *Winnipeg* cited Forsythe (1987) who had argued that the born-alive rule was an evidentiary standard based on "primitive medical science" (Forsythe 1987: 564) of a bygone era.
92 *Winnipeg, supra* n. 68 at 981 (S.C.C.).
93 Ibid. at 984.
94 Ibid.
95 Ibid. at 984–985.
96 Ibid. at 985.
97 Ibid. at 985, citing *Daigle*, *supra* n. 2.
98 Ibid. at 950.
99 Ibid.
100 Ibid. at 984.
101 Ibid. at 950.
102 Ibid.
103 Ibid. at 966.
104 Ibid. at 938, citing *Daigle*, *supra* n. 2.
105 See *Dehler v. Ottawa Civic Hospital* (1979), 25 O.R. 9 (2d) 748 at 761 (Ont. H.C.); *Borowski*, *supra.* n. 1 at 738–739 (Sask. C.A.).
106 *Winnipeg, supra* n. 68 (S.C.C.) at 953.
107 Ibid.
108 Ibid. at 959.
109 Ibid.
110 Ibid. at 978.
111 *Dobson (Litigation Guardian of) v. Dobson*, [1999] 2 S.C.R. 753. Hereinafter *Dobson.*
112 Ibid. at 768.
113 Ibid. at 770.
114 Ibid.
115 Ibid.
116 Ibid. at 771.
117 Ibid. at 784.
118 Ibid. at 785.
119 Ibid. at 784. The judgement itself placed "lifestyle choices" in quotations.
120 *Congenital Disabilities (Civil Liabilities) Act, 1976* (UK), 1976. For background to this act, see United Kingdom, Law Commission 1973 and 1974.
121 I thank my colleague Brian Singer for calling my attention to the importance of the symbolic in sovereignty. Our joint work (Singer and Weir forthcoming) examines the relation between sovereignty and governance.

Bibliography

A Regionalized System for Reproductive Medical Care in Ontario: Report to the Advisory Committee on Reproductive Medical Care to the Minister of Health for Ontario (1979) Ontario: Advisory Committee on Reproductive Medical Care to the Minister of Health.

Ackerknecht, E.H. (1976) "Midwives as Experts in Court", *Bulletin of the New York Academy of Medicine*, 10: 1224–1228.

Adair, F.L. and Edith Potter (1939) "Fetal and Neonatal Disease and Death", *American Journal of Obstetrics and Gynecology*, 37 (June): 993–999.

Adair, F.L. and Edith Potter (1936) "Fetal and Neonatal Mortality with Recommendations for Reduction", *American Journal of Public Health*, 26 (March): 281–286.

Adams, Mark (ed.) (1990) *The Wellborn Science: Eugenics in Germany, France, Brazil, and Russia*, New York and Oxford: Oxford University Press.

Adelswärd, Viveka and Lisbeth Sachs (1998) "Risk Discourse: Recontextualization of Numerical Values in Clinical Practice", *Text*, 18, 2: 191–210.

Advisory Committee on Reproductive Care (1988) *Reproductive Care: Towards the 1990s: Second Report of the Advisory Committee on Reproductive Care*, Toronto: Queen's Printer.

Advisory Committee on Reproductive Medical Care (1979) *A Regionalized System for Reproductive Medical Care in Ontario*, Toronto: Ontario Ministry of Health.

Agamben, Giorgio (2002) [1999] *Remnants of Auschwitz*, trans. D. Heller-Roazen, New York: Zone.

Agamben, Giorgio (1998) [1995] *Homo Sacer: Sovereign Power and Bare Life*, trans. D. Heller-Roazen, Stanford, CA: Stanford University Press.

Alexander, Sophie and Marc J.N.C. Keirse (1989) "Formal Risk Scoring during Pregnancy", in Iain Chalmers, Murray Enkin and Marc J.N.C. Keirse (eds), *Effective Care in Pregnancy and Childbirth*, Oxford, New York and Toronto: Oxford University Press, Vol. 1: 345–365.

American Medical Association, *A Guide for the Study of Perinatal Mortality and Morbidity* (1959) Committee on Maternal and Child Care, Council in Medical Service, American Medical Association.

Andrew, Donna T. (1989) *Philanthropy and Police*, Princeton, NJ: Princeton University Press.

Anspach, Renée R. (1993) *Deciding Who Lives: Fateful Choices in the Intensive Care Nursery*, Berkeley: University of California Press.

Apgar, Virginia (1953) "Proposal for New Method of Evaluation of Newborn Infants", *Anesthesia & Analgesia*, 32, July–Aug.: 260–267.

Armstrong, David (1997) "The Social Space of Illness", in Dorothy Porter (ed.) *Social Medicine and Medical Sociology in the Twentieth Century*, Amsterdam and Atlanta GA: Rodopi: 165–174.

Armstrong, David (1995) "The Rise of Surveillance Medicine", *Sociology of Health and Illness*, 17: 393–410.

Armstrong, David (1986) "The Invention of Infant Mortality", *Sociology of Health and Illness*, 8: 211–232.

Armstrong, Elizabeth M. (2003) *Conceiving Risk, Bearing Responsibility: Fetal Alcohol Syndrome and the Diagnosis of Moral Disorder*, Baltimore, MD and London: Johns Hopkins.

Arney, William (1982) *Power and the Profession of Obstetrics*, Chicago: University of Chicago Press.

Arnup, Katherine (1994) *Education for Motherhood: Advice for Mothers in Twentieth-Century Canada*, Toronto: University of Toronto Press.

Avery, Mary Ellen (1992) "A 50-Year Overview of Perinatal Medicine", *Early Human Development*, 29: 43–50.

Baird, Dugald (1962) *Combined Textbook of Obstetrics and Gynaecology*, Edinburgh and London: E&S Livingstone.

Baird, Dugald (1960) "The Evolution of Modern Obstetrics", *Lancet*, 2 (Sept. 10): 557–564.

Baird, Dugald (1947) "Social Class and Foetal Mortality", *Lancet*, 253 (Oct. 11): 531–535.

Baird, Dugald (1945) "The Influence of Social and Economic Factors on Stillbirths and Neonatal Deaths", *Journal of Obstetrics and Gynaecology of the British Empire*, N.S. 52, 3: 217–234.

Baird, Dugald and A.M. Thomson (1954) "The Epidemiological Approach in Obstetrics", *Gynaecologia*, 138: 226–245.

Baird, Dugald, J. Walker and A.M. Thomson (1954) "The Causes and Prevention of Stillbirths and First Week Deaths", Part III, "A Classification of Deaths by Clinical Cause: The Effect of Age, Parity and Length of Gestation on Death Rates by Cause", *The Journal of Obstetrics and Gynaecology of the British Empire*, N.S. 61, 4: 43–448.

Baird, Dugald, A.M. Thomson and Ethel H.L. Duncan (1953) "The Causes and Prevention of Stillbirths and First Week Deaths," Part II, "Evidence from Aberdeen Clinical Records", *Journal of Obstetrics and Gynaecology of the British Empire*, N.S. 60, 1: 17–30.

Baird, Dugald, and J.F.B. Wyper (1941) "High Stillbirth and Neonatal Mortalities", *Lancet*, 241 (Nov. 29): 657–659.

"Baird, Dr. Dugald." http:// www.africa2000.com/ENDX/eug_babh.htm (accessed August 2001).

Baker, Jeffrey P. (2000) "The Incubator and the Medical Discovery of the Premature Infant", *Journal of Perinatology*, 5: 321–328.

Baker, Jeffrey P. (1996) *The Machine in the Nursery*, Baltimore, MD and London: Johns Hopkins.

Baker, Tom and Jonathan Simon (eds) (2002) *Embracing Risk: The Changing Culture of Insurance and Responsibility*, Chicago: University of Chicago Press.

Bala, Nicholas (1991) "Child and Family Policies for the 1990s: Justice Issues", in Laura Johnson and Richard Barnhorst (eds), *Children, Families and Family Policy in the 90s*, Toronto: Thompson: 105–134.

Ballantyne, J.W. (1922) "Ante-natal, Intra-natal, and Neo-natal Death", *British Medical Journal* 2 (Sept. 30): 583–588.

Ballantyne, J.W. (1902) *Manual of Antenatal Pathology and Hygiene*, Vol. 1. *The Foetus*, Edinburgh: William Green.

Ballantyne, J.W. (1901) "A Plea for a Pro-Maternity Hospital", *British Medical Journal*, 1 (April 6): 2101–2102.

"John William Ballantyne" (Obituary) (1923) *British Medical Journal*, 1 (Feb. 3): 213–215.

Barclay, Alfred E. (1944) *The Foetal Circulation and Cardiovascular System, and the Changes They Undergo at Birth*, Oxford: Blackwell.

Barcroft, Sir Joseph (1946) *Researches on Pre-Natal Life*, Oxford: Blackwell.

BCMCF (British Columbia, Ministry for Children and Families) (1996) *The Risk Assessment Model for Child Protection in British Columbia*, Victoria: Queen's Printer.

Beal, Ron (1984) "'Can I Sue Mommy?' An Analysis of a Woman's Tort Liability for Prenatal Injuries to her Child Born Alive", *San Diego Law Review*, 21, 2: 325–370.

Beck, Ulrich (1995) "Risk Society and the Provident State," in Scott Lash, Bronislow Szerszynski and Brian Wynne (eds) *Risk, Environment and Modernity*, London: Sage: 27–43.

Beck, Ulrich (1992) *Risk Society: Towards a New Modernity*, London: Sage.

Bendix, Reinhard (1977) [1960] *Max Weber: An Intellectual Portrait*, Berkeley: University of California Press.

Benoit, Cecilia (1994) "Paradigm Conflict in the Sociology of Service Professions: Midwifery as a Case Study", *Canadian Journal of Sociology*, 19, 3: 303–329.

Benoit, Cecilia (1991) *Midwives in Passage: The Modernisation of Maternity Care*, St. John's, Newfoundland: Institute of Social and Economic Research.

Berg, Marc (1995) "Turning a Practice into a Science: Reconceptualizing Postwar Medical Practice", *Social Studies of Science*, 25: 437–476.

Berg, Marc and Annemarie Mol (eds) (1998) *Differences in Medicine*, Durham, NC and London: Duke University Press.

Biggs, Lesley (2004) "Rethinking the History of Midwifery in Canada", in Ivy Lynn Bourgeault, Cecilia Benoit, Robbie Davis-Floyd (eds), *Reconceiving Midwifery*, Montreal: McGill-Queen's: 17–45.

Biggs, Leslie (1983) "The Case of the Missing Midwives: A History of Midwifery in Ontario from 1795–1900", *Ontario History*, 75, 1: 21–35.

Birks, Peter (1986) "Fictions Ancient and Modern", in Neil MacCormick and Peter Birks (eds), *The Legal Mind: Essays for Tony Honoré*, Oxford: Clarendon Press: 83–101.

Bock, Gisela and Pat Thane (eds) (1991) *Maternity and Gender Policies: Women and the Rise of the European Welfare States 1880s–1950s*, London and New York: Routledge.

Bonar, Barnet E. (1935) "The Newborn", *Journal of Pediatrics*, 7, 1: 99–117.

Bonar, Barnet E. (1934) "Problems of the Newborn", *Journal of Pediatrics*, 5, 2: 216–243.

Bonar, Barnet E. (1932) "Problems of the Newborn", *Journal of Pediatrics*, 1, 1: 92–109.

Bound, J.P., N.R. Butler and W.G. Spector (1956a) "Classification and Causes of Perinatal Mortality", *British Medical Journal* (Dec. 1): 1260–1265.

Bound, J.P., N.R. Butler and W.P. Spector (1956b) "Classification and Causes of Perinatal Mortality", *British Medical Journal* (Nov. 24): 1191–1196.

Bourgeault, Ivy Lynn (2000) "Delivering the 'New' Canadian Midwifery: The Impact on Midwifery of Integration into the Ontario Health Care System", *Sociology of Health and Illness*, 23, 5: 633–63.

Bourgeault, Ivy Lynn (1996) "Delivering Midwifery: An Examination of the Process and Outcome of the Incorporation of Midwifery in Ontario", unpublished Ph.D. thesis, University of Toronto.

Bourgeault, Ivy Lynn, Cecilia Benoit and Robbie Davis-Floyd (eds) (2004a) *Reconceiving Midwifery*, Montreal: McGill-Queen's.

Bourgeault, Ivy Lynn, Cecilia Benoit and Robbie Davis-Floyd (2004b) "Reconceiving Midwifery in Canada", in Ivy Lynn Bourgeault, Cecilia Benoit and Robbie Davis-Floyd (eds), *Reconceiving Midwifery*, Montreal: McGill-Queen's: 3–14.

Bower, Lisa C. (1995) "The Trope of the Dark Continent in the Fetal Harm Debates: 'Africanism' and the Right to Choice", in Patricia Boling (ed.), *Expecting Trouble: Surrogacy, Fetal Abuse and New Reproductive Technologies*, Boulder, CO: Westview: 142–155.

Boyd, Susan C. (1999) *Mothers and Illicit Drugs*, Toronto: University of Toronto Press.

Brosco, Jeffrey (1999) "The Early History of the Infant Mortality Rate in America: A Reflection Upon the Past and a Prophesy for the Future", *Pediatrics*, 103, 2: 478–485.

Browne, Francis J. (1922) "Neo-Natal Death", *British Medical Journal*, 2 (Sept. 30): 590–593.

Browne, Francis J. (1921) "Stillbirth: Its Causes, Pathology and Prevention", *British Medical Journal* (July 30), 140–145.

Buchanan, Allan, D.W. Brock, N. Daniels and D. Winkler (2000) *From Chance to Choice: Genetics and Justice*, Cambridge, Cambridge University Press.

Buekens P. (1995) "Variations in Provision and Uptake of Antenatal Care", in E. Steegers, T. Estes and E. Symond (eds), *Preventative Care in Obstetrics*, London: Bailliere Tindall, 3rd edn: 187–205.

Bundesen, H.N., William I. Fishbein, O.A. Darms and E.L. Potter (1938) "Factors in Neonatal Deaths", *Journal of the American Medical Association*, 111: 134–141.

Bunton, Robin, Sarah Nettleton and Roger Burrows (1995) *The Sociology of Health Promotion: Critical Analyses of Consumption, Lifestyle and Risk*, London and New York: Routledge.

Burtch, Brian (1994) *Trials of Labour: The Re-emergence of Midwifery*, Montreal: McGill-Queen's.

Butler, Neville R. (1966) "The Causes and Prevention of Perinatal Mortality in the European Region", World Health Organisation, Regional Committee for Europe, Technical Discussion, Sixteenth Session, Rabat.

Butler, Neville R. and Eva D. Alberman (1969) *Perinatal Problems: The Second Report of the 1958 British Perinatal Mortality Survey*, Edinburgh and London: E&S Livingstone Ltd.

Butler, Neville R. and Dennis G. Bonham (1963) *Perinatal Mortality: The First Report of the 1958 British Perinatal Mortality Survey*, Edinburgh and London: E&S Livingstone Ltd.

Callahan, Marilyn (2000) "An Analysis of the Supreme Court Decision in the Case between Winnipeg Child and Family Services and Ms. G", in Deborah Rutman, Marilyn Callahan, Audrey Lundquist, Suzanne Jackson and Barbara Field, *Substance Use and Pregnancy: Conceiving Women in the Policy-Making Process*, Ottawa: Status of Women Canada: 25–46.

Campbell, Nancy (1999) "Regulating 'Maternal Instinct': Governing Mentalities of Late Twentieth-Century U.S. Illicit Drug Policy", *Signs* 24, 4: 895–922.

Canadian Task Force on the Periodic Health Examination (1994) *The Canadian Guide to Clinical Preventive Health Care*, Ottawa: Health Canada.

Canguilhem, Georges (1989) [1978] *The Normal and the Pathological*, trans. Carolyn R. Fawcett, New York: Zone.

Carroll, Dena and Cecilia Benoit (2004) "Aboriginal Midwifery in Canada: Merging Traditional Practices and Modern Science", in Ivy Lynn Bourgeault, Cecilia Benoit and Robbie Davis-Floyd (eds), *Reconceiving Midwifery*, Montreal: McGill-Queen's: 263–287.

Carroll, Jane *et al.* (1994) "Psychosocial Risk Factors during Pregnancy: What do Family Physicians Ask About?", *Canadian Family Physician*, 40: 1280–1289.

Casper, Monica J. (1998) *The Making of the Unborn Patient: A Social Anatomy of Fetal Surgery*, New Brunswick, NJ and London: Rutgers University Press.

Castel, Robert (1991) "From Dangerousness to Risk", in Graham Burchell, Colin Gordon and Peter Miller (eds), *The Foucault Effect*, Chicago: University of Chicago Press: 281–298.

Castel, Robert (1981) *La Gestion des risques; de l'anti-psychiatrie à l'après-psychanalyse*, Paris: Les Éditions de Minuit.

Cattell, Jacques (ed.) (1975) *American Men and Women of Science: The Medical Sciences*, New York and London: R.R. Bowker Company.

Causes of Stillbirth, 1943–55 (1957) Canada. Dominion Bureau of Statistics, Health and Welfare Division, Vital Statistics Section. Ottawa: Edmond Cloutier, Queen's Printer.

Chamallas, Martha with Linda Kerber (1990) "Women, Mothers, and the Law of Fright: A History", *Michigan Law Review*, 88, 4: 814–864.

Chance, Graham (1992) "Changes to the Ontario Antenatal Record", *Ontario Medical Review* (Dec.): 23–25.

Chatworth, Susan (1996) "Indications for Transfer for Childbirth in Women Served by the Innuulisivik Maternity", unpublished M.Sc. thesis, Department of Epidemiology and Biostatistics, McGill University, Montreal.

Chen, Xiaobei (2003) "Constituting 'Dangerous Parents' through the Spectre of Child Death: A Critique of Child Protection Restructuring in Ontario", in Deborah Brock (ed.), *Making Normal: Social Regulation in Canada*, Toronto: Thompson: 209–234.

Chen, Xiaobei (2000) "Is It all Neo-Liberal? Some Reflections on Child Protection Policy and Neo-Conservatism in Ontario", *Canadian Review of Social Policy*, 45–46: 237–246.

Clarren, S.K. and D.W. Smith (1978) "Fetal Alcohol Syndrome", *New England Journal of Medicine*, 298, 19: 1063–1067.

Clifford, Stewart (1936) "A Study of Neonatal Mortality", *Journal of Pediatrics*, 8: 367–374.

College of Midwives of Ontario (1994a) "Philosophy of Midwifery Care in Ontario", Toronto: College of Midwives (Jan.).

College of Midwives of Ontario (1994b) "Standard on Records Content", Toronto: College of Midwives (May 9), unpublished.

Comacchio, Cynthia (1993) *"Nations Are Built of Babies" : Saving Ontario's Mothers and Children, 1900–1940*, Montreal: McGill-Queen's, unpublished.

Comité d'étude de la mortalité périnatale (1967 and 1969) *Rapport du Comité d'étude de la mortalité périnatale du Québec*, Quebec: Comité d'étude de la mortalité périnatale, Parts 1 and 2.

Conaghan, Joanne and Wade Mansell (1993) *The Wrongs of Tort*, London and Boulder, CO: Pluto Press.

Condon, Mary (2001) "Of Butterflies and Bitterness?: Legal Fictions in Corporate and Securities Law", in Ysolde Gendreau (dir.), *Les fictions du droit – Fictions in the Law*. Montreal: Les Éditions Thémis: 125–145.

Conklin, Beth A. and Lynn M. Morgan (1996) "Babies, Bodies, and the Production of Personhood in North America and a Native Amazonian Society", *Ethos*, 24, 4: 657–694.

Connor, J.T.H. (1994) " 'Larger Fish to Catch Here than Midwives': Midwifery and the Medical Profession in Nineteenth-Century Ontario", in Diane Dodd and Deborah Gorham (eds), *Caring and Curing: Historical Perspectives on Women and Healing in Canada*, Ottawa: University of Ottawa Press: 103–134.

Corbett, Edward P.J. (1971) *Classical Rhetoric for the Modern Student*, New York: Oxford University Press. 2nd edn.

Corwin, E.H.L. (1952) *Infant and Maternal Care in New York City: A Study of Hospital Facilities*, New York: Columbia.

Cosbie, W.G. (1992) "The Obstetrical Causes and Prevention of Stillbirth and Early Infant Mortality", *Canadian Medical Association Journal*, 146, 7: 1203–1207 (reprint of Cosbie 1923).

Cosbie, W.G. (1923) "The Obstetrical Causes and Prevention of Stillbirth and Early Infant Mortality", *Canadian Medical Association Journal*, 13, 12: 877–880.

Cox, Susan M. and William McKellin (1999) "'There's This Thing in Our Family': Predictive Testing and the Construction of Risk for Huntington Disease", in Peter Conrad and Jonathan Gabe (eds), *Sociological Perspectives on the New Genetics*, Oxford: Blackwell: 121–145.

Crosse, V. Mary and Jean M. Mackintosh (1954) "Perinatal Mortality", in D. Gairdner (ed.), *Recent Advances in Paediatrics*, London: Churchill: 63–86.

Cruickshank, J.N. (1930) *Child Life Investigations: The Causes of Neo-natal Death*, Medical Research Council SRS No. 145. London: HMSO.

Cruickshank J.N. (1922) "Syphilis as a Cause of Ante-natal Death", *British Journal of Medicine* (Dec. 30), 3220: 593–594.

Cruikshank, Barbara (1999) *The Will to Empower: Democratic Citizens and Other Subjects,* Ithaca, NY: Cornell University Press.

Curtis, Bruce (2002) "Foucault on Governmentality and Population: The Impossible Discovery", *Canadian Journal of Sociology*, 27, 4: 505–533.

Curtis, Bruce (2001) *The Politics of Population: State Formation, Statistics, and the Census of Canada, 1840–1875*, Toronto: University of Toronto Press.

Curtis, Bruce (1995) "Taking the State Back Out: Rose and Miller on Political Power", *British Journal of Sociology*, 46, 4: 575–589.

Daniels, Cynthia (1997) "Between Fathers and Fetuses: The Social Construction of Male Reproduction and the Politics of Fetal Harm", *Signs*, 22, 31: 579–616.

Daniels, Cynthia (1993) *At Women's Expense: State Power and the Politics of Fetal Rights*, Cambridge, MA: Harvard University Press.

Danforth, David N. (1966) *Textbook of Obstetrics and Gynecology*, New York and London: Hoeber Medical Division, Harper and Row.

Davin, Anna (1978) "Imperialism and Motherhood", *History Workshop Journal*, 5: 3–66.

Dawber, T.R. (1980) *The Framingham Study: The Epidemiology of Coronary Heart Disease*, Cambridge, MA: Harvard University Press.

Dawson, Brettel (1991) "A Feminist Response to 'Unborn Child Abuse: Contemplating a Legal Solution'", *Canadian Journal of Family Law*, 9: 157–176.

Dawson, Brettel (1990) "*Re Baby R*: A Comment on Fetal Apprehension", *Canadian Journal of Women and the Law*, 4: 265–275.

Dean, Mitchell (2002) "Liberal Government and Authoritarianism", *Economy and Society*, 31, 1: 37–61.

Dean, Mitchell (1999a) *Governmentality: Power and Rule in Modern Society*, London: Sage.

Dean, Mitchell (1999b) "Risk, Calculable and Incalculable", in Deborah Lupton (ed.), *Risk and Sociocultural Theory: New Directions and Perspectives*, Cambridge: Cambridge University Press: 131–159.

Dean, Mitchell (1996) "Putting the Technological into Government", *History of the Human Sciences*, 9, 3: 47–68.

Deleuze, Gilles (1988) [1986] *Foucault*, trans. Sean Hand, Minneapolis: University of Minnesota Press.

Desmond, Murdina (1991) "A Review of Newborn Medicine in America: European Past and Guiding Ideology", *American Journal of Perinatology*, 8, 5: 308–322.

Detweiler, H.K. (1926) "A Survey of the Results of Routine and Other Wasserman Tests in the Toronto General Hospital for the Past Ten Years", *Canadian Public Health Journal*, 17: 543–550.

Dickens, Bernard (1989) "Wrongful Birth and Life, Wrongful Death before Birth, and Wrongful Law", in Sheila McLean (ed.), *Legal Issues in Human Reproduction*, Aldershot: Gower.

Diduck, Alison (1993) "Legislating Ideologies of Motherhood", *Social and Legal Studies*, 2: 461–485.

The Digest of Justinian (1985) Theodor Mommsen and Paul Krueger (eds), trans. Alan Watson, Philadelphia: University of Pennsylvania Press.

Dineen, Claire E. (1994) "Fetal Alcohol Syndrome: The Legal and Social Responses to its Impact on Native Americans", *North Dakota Law Review*, 70, 1: 1–66.

"Discussion on Stillbirth and Neonatal Mortality" (1942–1943) *Proceedings of the Royal Society of Medicine, Section of Obstetrics and Gynaecology* XXXVI: 59–68.

Doggett, Maeve Elizabeth (1988) "Coverture, the Fiction of Marital Unity and the Status of Wives: Legal Responses to Wife-Assault in Nineteenth-Century England in Context", unpublished LL.M. thesis, Osgoode Hall, York University, Canada.

Doran, Nob (1996) "From Embodied 'Health' to Official 'Accident': Class, Codification and British Factory Legislation, 1831–1844", *Social and Legal Studies*, 5, 4: 523–546.

Douglas, Mary (1992) *Risk and Blame: Essays in Cultural Theory*, London and New York: Routledge.

Dreyfus, Hubert and Paul Rabinow (1983) *Michel Foucault: Beyond Structuralism and Hermeneutics*, Chicago: University of Chicago Press.

Duden, Barbara (2002a) *Die Gene im Kopf – der Fötus in Bauch: Historisches zum Frauenkörper*, Hanover: Offizin.

Duden, Barbara (2002b) "Zwischen 'wahren Wissen' und Prophetie: Konzeptionen des Ungeborenen", in Barbara Duden, Jürgen Schlumbohn and Patrice Veit (eds), *Geschichte des Ungeborenen: zur Ehrfahrungs- und Wissenschaftsgeschichte der Schwangerschaft, 17–20 Jahrhundert*, Göttingen: Vandenhoeck and Ruprecht: 1–47.

Duden, Barbara (2000) "Hoffnung, Ahnung, 'sicheres Wissen': Zur Historisierung des Wissengrundes vom Schwangergehen", *Die Psychotherapeutin*, 13, S: 25–37.

Duden, Barbara (1999) "'The Fetus on the Farther Shore': Toward a History of the Unborn", in Lynn M. Morgan and Meredith W. Michaels (eds), *Fetal Subjects: Feminist Positions*, Philadelphia: University of Pennsylvania Press: 13–25.

Duden, Barbara (1993) *Disembodying Women*, trans. Lee Hoinacki, Cambridge, MA: Harvard University Press.

Duden, Barbara (1992) "Quick with Child: An Experience that Has Lost its Status", *Technology in Society*, 14: 335–344.

Duden, Barbara (1991) [1987] *The Woman beneath the Skin: A Doctor's Patients in Eighteenth-Century Germany*, trans. Thomas Dunlap, Cambridge, MA and London: Harvard University Press.

Duden, Barbara, Jürgen Schlumbohn and Patrice Veit (eds) (2002) *Geschichte des Ungeborenen: zur Ehrfahrungs- und Wissenschaftsgeschichte der Schwangerschaft, 17–20 Jahrhundert*, Göttingen: Vandenhoeck and Ruprecht.

Duncan, Ethel H.L., D. Baird and A.M. Thomson (1952) "The Causes and Prevention of Stillbirths and First Week Deaths", Part I, "The Evidence of Vital Statistics", *Journal of Obstetrics and Gynaecology of the British Empire*, N.S. 59, 2: 183–196.

Duncan, J.H. (1937) "Neo-Natal Mortality", *Canadian Medical Association Journal* 37, 5: 474–478.

Dunn, Peter M. (1993) "Hippocrates (460–*c*.356 BC) and the Founding of Perinatal Medicine", *Archives of Disease in Childhood*, 69: 1337–1352.

Dunstan, G.R. (1990) "Introduction: Text and Context", in G.R. Dunstan (ed), *The Human Embryo: Aristotle and the Arabic and European Tradition*, Exeter: University of Exeter Press: 1–9.

Dwork, Deborah (1987) *War Is Good for Babies and Other Young Children: A History of the Infant and Child Welfare Movement in England, 1898–1918*, London: Tavistock.

Ebbs, J.H., W.A. Scott, F.F. Tisdall, W.J. Moyle and M. Bell (1942) "Nutrition in Pregnancy", *Canadian Medical Association Journal*, 46, 1: 1–8.

Ebbs, J.H., F.F. Tisdall and W.A. Scott (1941) "The Influence of Prenatal Diet on the Mother and Child", *Journal of Nutrition*, 22: 515–526.

Eccles, Audrey (1982) *Obstetrics and Gynaecology in Tudor and Stuart England*, London: Croom Helm.

Eden, Thomas Watts (1931) "The Infant Mortality of Birth", *Lancet*, 221 (Dec. 5): 1223–1227.

Eden, Thomas Watts and Eardley Lancelot Holland (1937) *A Manual of Obstetrics*, London: J & A Churchill Ltd., 8th edn.

Edwards, J., Sarah Franklin, Eric Hirsch, Frances Price and Marilyn Strathern (1993) *Technologies of Procreation*, Manchester: Manchester University Press.

Enke, Ulrike (2002) "Von der Schönheit der Embryonenen: Samuel Thomas Soemmerrings Werk *Icones embryonum humanorum* (1799)", in Barbara Duden, Jürgen Schlumbohn and Patrice Veit (eds), *Geschichte des Ungeborenen: zur Ehrfahrungs- und Wissenschaftsgeschichte der Schwangerschaft, 17–20 Jahrhundert*, Göttingen: Vandenhoeck and Ruprecht: 205–236.

Ericson, Richard V. and Aaron Doyle (2004) *Uncertain Business: Risk, Insurance, and the Limits of Knowledge*, Toronto: University of Toronto Press.

Ericson, Richard V., Aaron Doyle and Dean Barry (2003) *Insurance as Governance*, Toronto: University of Toronto Press.

Evans, Patricia and Karen Swift (2000) "Single Mothers and the Press: Rising Tides, Moral Panic, and Restructuring Discourses", in Sheila M. Neysmith (ed.),

Restructuring Caring Labour: Discourse, State Practice, and Everyday Life, Toronto: Oxford University Press: 73–92.

Evans, R.G., Morris L. Barer and Theodore R. Marmor (1994) *Why Are Some People Healthy and Others Not? The Determinants of Health of Populations*, New York: Aldine de Gruyter.

Ewald, François (2002) "The Return of Descartes's Malicious Demon: An Outline of a Philosophy of Precaution", in Tom Baker and Jonathan Simon (eds), *Embracing Risk: The Changing Culture of Insurance and Responsibility*, Chicago: University of Chicago Press: 273–301.

Ewald, François (1991) "Insurance and Risk", in Graham Burchell, Colin Gordon and Peter Miller (eds), *The Foucault Effect*, Chicago: University of Chicago Press: 197–210.

Ewald, François (1990) "Norms, Discipline and the Law", *Representations*, 30: 138–161.

Ewald, François (1986) *L'État providence*, Paris: Grasset.

Feeley, Malcolm and Jonathan Simon (1994) "Actuarial Justice: The Emerging New Criminal Law", in David Nelken (ed), *The Futures of Criminology*, London: Sage: 173–201.

Feeley, Malcolm and Jonathan Simon (1992) "The New Penology: Notes on the Emerging Strategy of Corrections and its Implications", *Criminology*, 30 (4): 449–474.

Forsythe, Clarke D. (1987) "Homicide of the Unborn Child: The Born Alive Rule and Other Legal Anachronisms", *Valparaiso University Law Review*, 21: 563–629.

Foucault, Michel (2003) *"Society Must Be Defended": Lectures at the Collège de France, 1975–1976*, trans. David Macey, New York: Picador (translation of Foucault 1997).

Foucault, Michel (1998) "Foucault", in P. Rabinow, (series ed.) *Essential Works of Foucault 1954–1984*. J. B. Faubion (ed.) Vol. 2, *Aesthetics, Method, and Epistemology*, New York: The New Press: 459–463.

Foucault, Michel (1997a) "Governmentality", in P. Rabinow, (series ed.) *Essential Works of Foucault 1954–1984*. J. B. Faubion (ed.) Vol. 3, *Power*, New York: The New Press: 201–222.

Foucault, Michel (1997b) *"Il faut défendre la société": Cours au Collège de France, 1975–1976*, Paris: Gallimard/Seuil.

Foucault, Michel (1997c) "'*Omnes et Singulatim*': Toward a Critique of Political Reason", in P. Rabinow, (series ed.) *Essential Works of Foucault 1954–1984*. J. B. Faubion (ed.) Vol. 3, *Power*, New York: The New Press: 298–325.

Foucault, Michel (1997d) "The Politics of Health in the Eighteenth Century", in P. Rabinow, (series ed.) *Essential Works of Michel Foucault 1954–1984*. J.B. Faubion (ed.) Vol. 3: *Power*, New York: The New Press: 90–104.

Foucault, Michel (1997e) "Truth and Power", in P. Rabinow, (series ed.) *Essential Works of Foucault 1954–1984*. J. B. Faubion (ed.) Vol. 3, *Power*, New York: The New Press: 111–132.

Foucault, Michel (1997f) "What Is Enlightenment?", in P. Rabinow, (series ed.) *Essential Works of Foucault 1954–1984*. Paul Rabinow (ed.) Vol. 1, *Ethics*, New York: The New Press: 303–320.

Foucault, Michel (1989) [1966] *The Order of Things*, London and New York: Routledge.

Foucault, Michel (1988a) *Résumé de cours*, Paris: Julliard.

Foucault, Michel (1988b) "On Problematization", *History of the Present*, Spring: 16–17.

Foucault, Michel. (1979) [1975] *Discipline and Punish: The Birth of the Prison*, trans. A. Sheridan, New York: Vintage.

Foucault, Michel (1978) [1976] *The History of Sexuality*, Vol. 1: *An Introduction*, trans. R. Hurley, New York: Pantheon.

Foucault, Michel (1972) [1969 and 1971] *The Archaeology of Knowledge and the Discourse on Language*, trans. A. Sheridan, New York: Harper Colophon (translation of Foucault 1969 and 1971).

Foucault, Michel (1963) *Naissance de la clinique*, Paris: Presses Universitaires de France.

Franklin, Sarah (1991) "Fetal Fascinations: New Dimensions to the Medical-Scientific Construction of Fetal Personhood", in S. Franklin, C. Lury and J. Stacey (eds), *Off-Centre: Feminism and Cultural Studies*, London: HarperCollins: 190–206.

Frohock, Fred (1986) *Special Care: Medical Decisions at the Beginning of Life*, Chicago: University of Chicago Press.

Fudge, Judy and Eric Tucker (1993) "Reproductive Hazards in the Workplace: Legal Issues of Regulation, Enforcement, and Redress", in Royal Commission on New Reproductive Technologies (Canada), Research Volumes, Vol. 4: *Legal and Ethical Issues in New Reproductive Technologies: Pregnancy and Parenthood*, Ottawa: Ministry of Supply and Services, 1994: 161–338.

Fuller, Lon (1967) *Legal Fictions*. Stanford, CA: Stanford University Press.

Gabe, Jonathan (ed.) (1995) *Medicine, Health and Risk*, Oxford: Blackwell.

Gavigan, Shelley (1984) "The Criminal Sanction as it Relates to Human Reproduction: The Genesis of the Statutory Prohibition of Abortion", *The Journal of Legal History*, 5, 1: 20–41.

Gélis, Jacques (1991) *History of Childbirth: Fertility, Pregnancy and Birth in Early Modern Europe*, trans. R. Morris, Cambridge: Polity Press.

Gifford, Sandra M. (1986) "The Making of Lumps: A Case Study of the Ambiguities of Risk", in Craig R. Janes, Ron Stall and Sandra Gifford (eds), *Anthropology and Epidemiology*, Dordrecht: Reidel: 213–246.

Gilbert-Barness, Enid (1995) "Edith Potter", *Journal of Obstetrics*, 126, 5, Pt. 1: 845–846.

Ginn, Diana (1999) "The Supreme Court of Canada Rules on Coercive State Intervention in Pregnancy", *Canadian Woman Studies*, 19, 1 and 2: 122–132.

Ginn, Diana E. (1994) "Pregnant Women and Consent to Medical Treatment", *Health Law in Canada*, 15, 2: 41–48.

Goffman, Erving (1997) "The Self as Ritual Object: The Nature of Deference and Demeanor", in Charles Lemert and Ann Branaman (eds), *The Goffman Reader*, Oxford: Blackwell: 27–34.

Goldstein, Jan (1984) "Foucault Among the Sociologists: The 'Disciplines' and the History of the Professions", *History and Theory*, 23: 170–192.

Gomez, Laura E. (1997) *Misconceiving Mothers: Legislators, Prosecutors and the Politics of Prenatal Drug Exposure*, Philadelphia: Temple University Press.

Goodrich, Peter (1990) *Languages of Law*, London: Weidenfeld and Nicholson.

Goodwin, James W. and Graham W. Chance (1979) "New System for Managing High-Risk Pregnancies", *Ontario Medical Review*, November: 563–567.

Goodwin, J.W, J.T. Dunne and B.W. Thomas (1969) "Antepartum Identification of the Fetus at Risk", *Canadian Medical Association Journal*, 101, 8: 57–67.

Gordon, Colin (1991) "Governmental Rationality: An Introduction", in Graham Burchell, Colin Gordon and Peter Miller (eds), *The Foucault Effect*, Chicago: University of Chicago Press: 1–52.

Gray-Withers, Diane (1997) "Decentralized Social Services and Self-Government: Challenges for First Nations", in Jane Pulkingham and Gordon Ternowetsky (eds),

Child and Family Policies: Struggles, Strategies and Options, Halifax, Nova Scotia: Fernwood: 85–99.

Greaves, Lorraine, Colleen Varcoe, Nancy Poole, Marina Morrow, Joy Johnson, Ann Pederson and Lori Irwin (2002) *A Motherhood Issue: Discourses on Mothering Under Duress*, Ottawa: Status of Women Canada.

Greschner, Donna (1900) "Abortion and Democracy for Women: A Critique of Tremblay v. Daigle" (Case comm.), *McGill Law Journal*, 35, May: 633–669.

Grulee, Clifford G. (1933) "Fetal and Neonatal Mortality", *Journal of Pediatrics*, 3, 1: 132–143.

Hacking, Ian (1990) *The Taming of Chance*, Cambridge: Cambridge University Press.

Haggerty, Kevin D. (2003) "From Risk to Precaution: The Rationalities of Personal Crime Prevention", in Richard V. Ericson and Aaron Doyle (eds), *Risk and Morality*, Toronto: University of Toronto Press: 193–214.

Hahn, Robert A. (1987) "Divisions of Labor: Obstetricians, Women, and Society in *Williams Obstetrics*, 1903–1985", *Medical Anthropology Quarterly*, New Series 1, 3: 256–282.

Hall, Philip (1994) "Rethinking Risk", *Canadian Family Physician*, 40: 1239–1244.

Halliday, M.A.K. (1985) *An Introduction to Functional Grammar*, London: Edward Arnold.

Hamilton, K. Scott (1989) "Prolegomenon to Myth and Fiction in Legal Reasoning, Common Law Adjudication and Critical Legal Studies", *The Wayne Law Review*, 35, 4: 1449–1480.

Hanigsberg, Julia E. (1991) "Power and Procreation: State Interference in Pregnancy", *Ottawa Law Review*, 231: 35–70.

Hannah, M.E., W.J. Hannah, S.A. Hewson, E.D. Hodnett, S. Saigal and A.R. Willan (2000) "Planned Caesarian Section versus Planned Vaginal Birth for Breech Presentation at Term: A Randomised Multicentre Trial", Term Breech Trial Collaborative Group, *Lancet*, 356 (Oct. 21): 1375–1383,

Hannah-Moffat, Kelly (1995) "Feminine Fortresses: Women-Centred Prisons?", *The Prison Journal*, 75, 2: 135–164.

Hanvey, L., P. Mohide, A. Shennan and N. Sutherland (1991) *Review of Maternal and Newborn Hospital Services in Ontario: Final Report of the Hospital Assessment Team*, Toronto: Ministry of Health, Ontario.

Harer, W. Benson (1932) "A Study of Fetal Mortality", *American Journal of Obstetrics and Gynecology*, 24, 2: 254–262.

Hartouni, Valerie (1997) *Cultural Conceptions: On Reproductive Technologies and the Remaking of Life*, Minneapolis: University of Minnesota Press.

Harvey, Sheila, Karyn Kaufman and Alison Rice (1995) "Hospital-Based Midwifery Projects in Canada", in Tricia Murphy-Black (ed.), *Issues in Midwifery*, Edinburgh: Churchill-Livingstone: 189–206.

Hayes, Michael V. and James R. Dunn (1998) "Population Health in Canada: A Systematic Review", Canadian Policy Research Network, Study No. H-01, Ottawa.

Health Canada (1996), Joint Statement by Health Canada, the Aboriginal Nurses of Canada, the Assembly of First Nations Health Commission, the Native Physicians Association in Canada, the Inuit Women's Association, "Presentation on Fetal Alcohol Syndrome (FAS) and Fetal Alcohol Effects (FAE) in Canada", October. Unpublished.

Hedgecoe, Adam (2004) *The Politics of Personalised Medicine: Pharmacogenetics in the Clinic*, Cambridge: Cambridge University Press.

Heidegger, Martin (1971) "Language", in Martin Heidegger, *Poetry, Language and Thought*, trans. Albert Hofstadter, New York: Harper Colophon: 187–210.

Heidegger, Martin (1959) *Unterwegs zur Sprache*, Tübingen: Neske.

Heimer, Carol A. and Lisa R. Staffen (1998) *For the Sake of the Children: The Social Organization of Responsibility in the Hospital and the Home*, Chicago and London: University of Chicago Press.

Herschkorn-Barnu, Paule (2002) "Wie der Fötus einen klinische Status erhielt: Bedingungen und Verfahren der Produktion eines medizinishcen Fachwissens, Paris 1832–1838", in Barbara Duden, Jürgen Schlumbohn and Patrice Veit (eds), *Geschichte des Ungeborenen: zur Ehrfahrungs- und Wissenschaftsgeschichte der Schwangerschaft, 17–20 Jahrhundert*, Göttingen: Vandenhoeck and Ruprecht: 167–204.

Hindess, Barry (2001) "The Liberal Government of Unfreedom", *Alternatives: Global, Local, Political*, 26, 2: 93–107.

History of the United States National Committee on Vital and Health Statistics, 1949–1964 (1966) Washington, D. C.: National Centre for Health Statistics, Public Health Service Publication No. 1000, Series 4, No. 5.

Hoffman, N.Y. (1982) "Edith Potter, M.D., Ph.D.: Pioneering Infant Pathology", *Journal of the American Medical Association*, 248, 13: 1551–1553.

Hofmeyer, J. and M. Hannah (2006) "Five Years to the Term Breech Trial: The Rise and Fall of a Randomized Controlled Trial", *American Journal of Abstetrics and Gynecology*, April 22 (Epub ahead of print).

Holland, Eardley L. (1922) *The Causation of Foetal Death*, London: HMSO.

Holland, Eardley L. and Janet E. Lane-Claypon (1926) *Child Life Investigations: A Clinical and Pathological Study of 1,673 Cases of Dead-Births and Neo-Natal Deaths*, Medical Research Council Special Report Series, No. 109, London: HMSO.

Holt, L. Emmett and Ellen C. Babbitt (1915) "Institutional Mortality of the New-Born", *Journal of the American Medical Association*, 64, 4: 287–290.

Hopwood, Nicholas (2002) "Embryonen 'auf dem Altar der Wissenschaft zu opfern': Entwicklungsreihen im späten neunzehten Jahrhundert", in Barbara Duden, Jürgen Schlumbohn and Patrice Veit (eds), *Geschichte des Ungeborenen: zur Ehrfahrungs- und Wissenschaftsgeschichte der Schwangerschaft, 17–20 Jahrhundert*, Göttingen: Vandenhoeck and Ruprecht: 237–272.

Hopwood, Nicholas (2000) "Producing Development: The Anatomy of Human Embryos and the Norms of Wilhelm His", *Bulletin of the History of Medicine*, 74, 1: 29–79.

Horwitz, M. (1977) *The Transformation of American Law, 1780–1860*, Cambridge, MA: Harvard University Press.

Hunt, Alan and Gary Wickham (1994) *Foucault and Law*, London: Pluto Press.

Hunter, David S. and Marc J.N.C. Keirse (1989) "Gestational Diabetes", in Iain Chalmers, Murray Enkin and Marc J.N.C. Keirse (eds), *Effective Care in Pregnancy and Childbirth*, Oxford, New York and Toronto: Oxford University Press, Vol. 1: 345–365.

Hunter, Ian (1994) *Rethinking the School: Subjectivity, Bureaucracy, Criticism*, New York: St. Martin's Press.

Ikemoto, Lisa C. (1992) "The Code of Perfect Pregnancy: At the Intersection of the Ideology of Motherhood, the Practice of Defaulting to Science, and the Interventionist Mindset of Law", *Ohio State Law Journal*, 53, 5: 1205–1306.

Irigaray, Luce (1985) *Speculum of the Other Woman*, trans. Gillian C. Gill, Ithaca, NY: Cornell University Press.

Jackman, Martha (1993) "The Canadian Charter as a Barrier to Unwanted Medical Treatment of Pregnant Women in the Interests of the Foetus", *Health Law in Canada*, 14, 2: 49–58.

Jackson, Beth E. (2005) *Practices of Knowing Population Health: A Study in Authorizing and Stabilizing Scientific Knowledge*, unpublished Ph.D. thesis, York University, Canada.

Jacobson, Joseph L. and Sandra W. Jacobson (2002) "Effects of Prenatal Alcohol Exposure on Child Development", *Alcohol Research and Health*, 26, 4: 282–286.

Jasanoff, Sheila (1995) *Science at the Bar: Law, Science, and Technology in America*, Cambridge, MA and London: Harvard University Press.

Jefferys, Margot (1997) "Social Medicine and Medical Sociology 1950–1970", in Dorothy Porter (ed.), S*ocial Medicine and Medical Sociology in the Twentieth Century*, Amsterdam and Atlanta, GA: Rodopi: 120–136.

Johnsen, Dawn E. (1986) "The Creation of Fetal Rights: Conflicts with Women's Constitutional Rights to Liberty, Privacy, and Equal Protection", *The Yale Law Journal*, 95: 599–625.

Johnston, Patrick (1983) *Native Children and the Child Welfare System*, Toronto: Lorimer.

Jones, K.L. and D.W. Smith (1973) "Recognition of the Fetal Alcohol Syndrome in Early Infancy", *Lancet*, 2 (Nov. 3) 2: 999–1001.

Jordanova, Ludmilla (1995) "Interrogating the Concept of Reproduction in the Eighteenth Century", in Faye D. Ginsburg and Rayna Rapp (eds), *Conceiving the New World Order: The Global Politics of Reproduction*, Berkeley: University of California Press: 369–386.

Kahn, Judith (1987) "Of Woman's First Disobedience: Forsaking a Duty of Care to Her Fetus – Is This a Mother's Crime?", *Brooklyn Law Review*, 53, 3: 807–843.

Kaufert, Patricia and John O'Neil (1993) "Analysis of a Dialogue on Risks in Childbirth: Clinicians, Epidemiologists and Inuit Women", in Shirley Lindenbaum and Margaret Lock (eds), *Knowledge, Power and Practice*, Berkeley: University of California Press: 32–54.

Kemshall, Hazel (2002) *Risk, Social Policy and Welfare*, Buckingham, UK, and Philadelphia: Open University Press.

Kemshall, Hazel (1998) *Risk in Probation Practice*, Aldershot: Ashgate.

Kerr, Ian R. (1997) "Pre-Natal Fictions and Post-Natal Actions", *Dalhousie Law Journal*, 20, 1: 237–274.

Kevles, Daniel J. (1995) *In the Name of Eugenics: Genetics and the Uses of Human Heredity*, Cambridge, MA, and London: Harvard University Press.

King, Helen (1990) "Making a Man: Becoming Human in Early Greek Medicine", in G.R. Dunstan (ed.), *The Human Embryo: Aristotle and the Arabic and European Tradition*, Exeter: University of Exeter Press: 10–19.

Klaus, Alisa (1993a) *Every Child a Lion: The Origins of Maternal and Infant Health Policy in the United States and France, 1890–1920*, Ithaca, NY: Cornell University Press.

Klaus, Alisa (1993b) "Depopulation and Race Suicide: Maternalism and Pronatalist Ideologies in France and the United States", in Seth Koven and Sonya Michel (eds), *Mothers of a New World: Maternalist Politics and the Origins of Welfare States*, New York: Routledge: 188–212.

Kline, Marlee (1993) "Complicating the Ideology of Motherhood: Child Welfare Law and First Nation Women", *Queen's Law Journal*, 18: 306–342.

Knight, Frank Hyneman (1921) *Risk, Uncertainty and Profit*, Boston: Houghton Mifflin.

Kohl, Stanley (1955) *Perinatal Mortality in New York City: Responsible Factors*, Cambridge, MA: Harvard University Press.

Kolder, V.E., J. Gallagher and M.T. Parsons (1987) "Court-Ordered Obstetrical Interventions", *New England Journal of Medicine*, 316, 19: 1192–1196.

Koven, Seth and Sonya Michel (eds), (1993) *Mothers of a New World: Maternalist Politics and the Origins of Welfare States*, New York: Routledge.

Koven, Seth and Sonya Michel (1990) "Womanly Duties: Maternalist Politics and the Origins of Welfare States in France, Germany, Great Britain and the United States, 1880–1920", *American Historical Review*, 95 (Oct. 4): 1067–1108.

Krieger, Nancy (2001) "Theories for Social Epidemiology in the Twenty-First Century: An Ecosocial Perspective", *International Journal of Epidemiology*, 30: 668–677.

Krieger, Nancy (1994) "Epidemiology and the Web of Causation: Has Anyone Seen the Spider?", *Social Science and Medicine*, 39, 7: 887–903.

Lancet (1980) "Sir Dugald Baird's Birthday Present", (Oct. 18): 839–840.

Lane, Karen (1995) "The Medical Model of the Body as a Site of Risk: A Case Study of Childbirth", in Jonathan Gabe (ed.), *Medicine, Health and Risk*, Oxford: Blackwell: 53–72.

Laprairie, Carol (1996) *Examining Aboriginal Corrections in Canada*, Ottawa: Minster of the Solicitor General.

Last, John M. (2001) *A Dictionary of Epidemiology*, New York: Oxford University Press. 4th edn.

League of Nations Health Organisation (1930) *Memorandum Relating to the Enquiries into the Causes and Prevention of Still-births and Mortality during the First Year of Life*, Geneva: Series of League of Nations Publications III, Health III, 7.

Lefort, Claude (1988) [1986] *Democracy and Political Theory*, Cambridge: Polity.

Lewis, Jane (1980) *The Politics of Motherhood: Child and Maternal Welfare in England, 1900–1939*, London: Croom Helm.

Lilienfeld, Abraham, Benjamin Pasamanick and Martha Rogers (1955) "Relationship between Pregnancy Experience and the Development of Certain Neuropsychiatric Disorders in Childhood", *American Journal of Public Health*, 45, 1: 637–643.

Lintgen, Charles A. (1962) "The Impact of Medical Knowledge on the Law Relating to Prenatal Injuries", *The University of Pennsylvania Law Review*, 110: 554–601.

Litt, Jacquelyn and Maureen McNeil (1997) "Biological Markers and Social Differentiation: Crack Babies and the Construction of the Dangerous Mother", *Health Care for Women International*, 18: 31–41.

Lock, Margaret (2002) *Twice Dead: Organ Transplants and the Reinvention of Death*, Berkeley: University of California Press.

Logan, Enid (1999) "The Wrong Race, Committing Crime, Doing Drugs, and Maladjusted for Motherhood: The Nation's Fury over 'Crack Babies' ", *Social Justice*, 26, 1: 115–138.

Lovell, P.A. and R.H. Griffith-Jones (1974) " 'The Sins of the Fathers' – Tort Liability for Pre-Natal Injuries", *The Law Quarterly Review*, 90: 531–558.

Luhmann, Niklas (1989) "Law as a Social System", *Northwestern Law Review*, 83: 136–150.

Lundquist, Audrey and Suzanne Jackson (2000) "An Aboriginal Nation's Efforts to Address Substance Use During Pregnancy", in Deborah Rutman, Marilyn Callahan, Audrey Lundquist, Suzanne Jackson and Barbara Field, *Substance Use and Pregnancy:*

Conceiving Women in the Policy-Making Process, Ottawa: Status of Women Canada: 139–154.

Lupton, Deborah (1999) "Risk and the Ontology of Pregnant Embodiment", in Deborah Lupton (ed.), *Risk and Sociocultural Theory: New Directions and Perspectives*, Cambridge: Cambridge University Press: 59–85.

Lupton, Deborah (1997) "Foucault and the Medicalisation Critique", in Alan Petersen and Robin Bunton (eds), *Foucault, Health and Medicine*, London: Routledge: 94–110.

Lupton, Deborah (1995) *The Imperative of Health*, London: Sage.

Lyon E.C. and G.C. Bemis (1931) "Study of Neonatal Deaths Occurring in 6000 Consecutive Deliveries", *American Journal of Obstetrics and Gynecology*, 21 (March): 373–380.

McCleary G.F. (1933) *The Early History of the Infant Welfare Movement*, London: H.K. Lewis.

McClive, Cathy (2002) "The Hidden Truths of the Belly: The Uncertainties of Pregnancy in Early Modern Europe", *Social History of Medicine*, 15, 2: 209–227.

McCormack, Thelma (1999) "Fetal Syndromes and the Charter: The Winnipeg Glue Sniffing Case", *Canadian Journal of Law and Society*, 14, 2: 77–101.

McIlroy, A. Louise (1922) "Foetal Death Due to Toxaemia", *British Medical Journal* (Sept. 20): 589–590.

McIntosh, C. Alison (1982) "The Rise of Twentieth-Century Pronatalism", *International Journal of Politics*, 12, 3: 42–80.

McLaren, Angus (1990) *Our Own Master Race: Eugenics in Canada, 1885–1945*, Toronto: McClelland and Stewart.

McLaren, Angus (1984) *Reproductive Rituals: The Perception of Fertility in England from the Sixteenth Century to the Nineteenth Century*, London and New York: Methuen.

MacMurchy, Helen (1926) "Neo-Natal Mortality", *The Public Health Journal* (Toronto), 16: 442–447.

McQuarrie, J.G. (1919) "Fetal Death: A Study of One Hundred and Nineteen Deaths in Series of Cases", *Journal of the American Medical Association*, 73, 21: 1574–1576.

Maier, Kelly E. (1989) "Pregnant Women: Fetal Containers or People with Rights?", *Affilia* 4, 2: 8–20.

Martin, David (1995) "The League of Nations Health Organization", in Paul Weindling (ed.), *International Health Organisations and Movements, 1918–1939*, Cambridge: Cambridge University Press: 56–80.

Martin, Sheilah and Murray Coleman (1995) "Judicial Intervention in Pregnancy", *McGill Law Journal*, 40: 947–991.

Massey, Elizabeth (1993) "By Her Own Authority: The Scope of Midwifery Practice Under the Ontario Midwifery Act, 1991", *Alberta Law Review*, XXXI, 1: 349–390.

"Maternal Syphilis and Fetal Mortality" (1920) *British Medical Journal*, (July 24): 134.

Mathews, Joan J. and Kathleen Zadak (1991) "The Alternative Birth Movement in the United States: History and Current Status", *Women and Health*, 17, 1: 39–56.

Means, Cyril Jr. (1968) "The Law of New York Concerning Abortion and the Status of the Foetus, 1664–1968: A Case of Cessation of Constitutionality", *New York Law Forum*, 14, 3: 411–515.

Meckel, Richard A. (1990) *Save the Babies: American Public Health and the Prevention of Infant Mortality, 1850–1929*, Baltimore, MD: Johns Hopkins.

Memmi, Dominique (1996) *Les Gardiens du corps*, Paris: Éditions de l'école des hautes études.

Michael, M. (1992) “Lay Discourses of Science: Science-in-general, Science-in-particular, and Self”, *Science, Technology and Human Values*, 17: 313–333.

Michaels, Meredith W. and Lynne M. Morgan (eds) (1999) *Fetal Subjects, Feminist Positions*, Philadelphia: University of Pennsylvania Press.

Miller, J.R. (1996) *Shingawauk's Vision: A History of Native Residential Schools*, Toronto: University of Toronto Press.

Miller, Peter and Nikolas Rose (1995) “Political Thought and the Limits of Orthodoxy: A Response to Curtis”, *British Journal of Sociology*, 46, 4: 590–597.

Miller, Peter and Nikolas Rose (1990) “Governing Economic Life”, *Economy and Society*, 19, 1: 1–31.

Milloy, John Sheridan (1999) *A National Crime: The Canadian Government and the Residential School System, 1879–1896*, Winnipeg: University of Manitoba Press.

Mitchell, Lisa M. (2001) *Baby's First Picture: Ultrasound and the Politics of Fetal Subjects*, Toronto: University of Toronto Press.

Mitchell, Lisa and Eugenia Georges (1997) “ ‘Baby's First Picture’: The Cyborg Fetus of Ultrasound”, *Feminist Studies* 23, 2: 373–401.

Mitchinson, Wendy (2002) *Giving Birth in Canada, 1900–1950*, Toronto: University of Toronto Press.

Monture-Angus, Patricia (1995) *Thunder in My Soul: A Mohawk Woman Speaks*, Halifax, Nova Scotia: Fernwood.

Mooney, Graham (1994) “Still-births and the Measurement of Urban Infant Mortality Rates *c.* 1890–1930”, *Local Population Studies*, 53: 42–54.

Morgan, Lynn (1996) “Fetal Relationality in Feminist Philosophy: An Anthropological Critique”, *Hypatia* 11, 3: 47–70.

Morgan, Lynn (1989) “When Does Life Begin? A Cross-Cultural Perspective on the Personhood of Fetuses and Young Children”, in Edd Doerr and James N. Prescott (eds), *Abortion Rights and Fetal “Personhood”*, Long Beach, CA: Certline Press for Americans for Religious Liberty. 2nd edn: 89–107.

Mosoff, Judith (1997) “ ‘A Jury Dressed in Medical White and Judicial Black’: Mothers with Mental Health Histories in Child Welfare and Custody”, in Susan B. Boyd (ed.), *Changing the Public/Private Divide: Feminism, Law and Public Policy*, Toronto: University of Toronto Press: 227–252.

Murphy, Douglas P. and James E. Bowman (1932) “The Frequency and Causes of Premature Birth”, *American Journal of Obstetrics and Gynecology*, 24, 2: 273–276.

Murphy, Patricia Aikins (1994) “Risk, Risk Assessment and Risk Labels”, *Journal of Nurse-Midwifery*, 39, 2: 67–69.

Mykhalovskiy, Eric and Lorna Weir (2004) “The Problem of Evidence-Based Medicine: Directions for Social Science”, *Social Science and Medicine*, 59: 1059–1069.

Nesbitt, Robert E.L., Jr. and Richard H. Aubrey (1969) “High-Risk Obstetrics”, *American Journal of Obstetrics and Gynecology*, 103: 972–985.

Newman, Karen (1996) *Fetal Positions: Individualism, Science, Visuality*, Stanford, CA: Stanford University Press.

Newsholme, Arthur (1910) *Thirty-ninth Annual Report, 1909–1910. Supplement to the Report of the Board's Medical Officer, Containing a Report on Infant and Child Mortality*. House of Commons Parliamentary Papers, Vol. XXXI: 973–1128. London.

Oakley, Ann (1991) “Eugenics, Social Medicine and the Career of Richard Titmuss in Britain 1935–50”, *British Journal of Sociology*, 42, 2: 165–194.

Oakley, Ann (1984) *The Captured Womb: A History of the Medical Care of Pregnant Women*, Oxford: Blackwell.

Oakley, Ann (1982) "The Origins and Development of Antenatal Care", in Murray Enkin and Iain Chalmers (eds), *Effectiveness and Satisfaction in Antenatal Care. Clinics in Developmental Medicine*, Nos. 81/82, Spastics International Medical Publications, London: William Heinemann Medical Books and Philadelphia: J.B. Lippincott: 1–21.

Oakley, Ann and Susanne Houd (1990) *Helpers in Childbirth: Midwifery Today*, Hemisphere Publishing Corporation for the World Health Organisation, Regional Office for Europe. Geneva.

Oaks, Laury (2000) "Smoke-Filled Wombs and Fragile Fetuses: The Social Politics of Fetal Representation", *Signs*, 26, 1: 63–108.

Oldham, J. (1985) "On Pleading the Belly: A History of Jury Matrons", *Criminal History*, 6: 1–64.

Oliver, Jean (1993/4) "State Intervention during Pregnancy and Childbirth", *Health Law Review*, 2, 3: 3–10.

Olivier, Pierre J.J. (1975) *Legal Fictions in Practice and Legal Science*, Rotterdam: Rotterdam University Press.

OMA Subcommittee on the Antenatal Record (2000) "A Guide to the Revised Antenatal Record of Ontario", *Ontario Medical Review*, 67, 3: 42–50.

O'Malley, K.D. and J. Nanson (2002) "Clinical Implications of a Link between Fetal Alcohol Spectrum Disorder and Attention-Deficit Hyperactivity Disorder", *Canadian Journal of Psychology*, 47, 4: 349–354.

O'Malley, Pat (2004) *Risk, Uncertainty and Government*, London: Cavendish.

O'Malley, Pat (2003) "Moral Uncertainties: Contract Law and Distinctions between Speculation, Gambling, and Insurance", in Richard V. Ericson and Aaron Doyle (eds), *Risk and Morality*, Toronto: University of Toronto Press: 231–257.

O'Malley, Pat (2000) "Introduction: Configurations of Risk", *Economy and Society*, 29, 4: 457–459.

O'Malley, Pat (1999) "Consuming Risks, Harm Minimisation and the Government of 'Drug Users'", in Russell Smandych (ed.), *Governable Places: Readings on Governmentality and Crime Control*, Aldershot: Dartmouth: 191–214.

O'Malley, Pat (1997) "Consuming Risks", in Mariana Valverde (ed.), *New Forms of Governance: Theory, Practice, Research*, Toronto: Centre of Criminology, University of Toronto: 29–30.

O'Malley, Pat (1996) "Risk and Responsibility", in Andrew Barry, Thomas Osborne and Nikolas Rose (eds), *Foucault and Political Reason*, Chicago: University of Chicago Press: 189–208.

O'Malley, Pat (1994) "Risk and Risk Management", in Mariana Valverde (comp.), *Radically Rethinking Regulation*, Toronto: Centre of Criminology, University of Toronto: 28–29.

O'Malley, Pat (1992) "Risk, Power and Crime Prevention", *Economy and Society*, 21, 3: 252–275.

O'Neil, John and Patricia Kaufert (1990) "The Politics of Obstetric Care: The Inuit Experience", in W. Penn Handwerker (ed.), *Births and Power: Social Change and the Politics of Reproduction*, Boulder, CO: Westview: 53–69.

The Ontario Council of Health (1971) *Perinatal Problems*, Ontario Department of Health, Monograph No. 2.

Ontario Perinatal Mortality Study Committee (1967a) *Second Report of the Perinatal Mortality Study in Ten University Teaching Hospitals, Ontario, Canada*, Toronto: Ontario Department of Health, Maternal and Child Health Service.

Ontario Perinatal Mortality Study Committee (1967b) *Second Report of the Perinatal Mortality Study in Ten University Teaching Hospitals, Ontario, Canada: Supplement*, Toronto: Ontario Department of Health, Maternal and Child Health Service.

Ontario Perinatal Mortality Study Committee (1961a) *Report of the Perinatal Mortality Study, (1959) in Ten University Teaching Hospitals, Ontario, Canada*, Section 1 – Summary, Toronto: Maternal and Child Health Branch, Department of Health.

Ontario Perinatal Mortality Study Committee (1961b) *Report of the Perinatal Mortality Study, (1959) in Ten University Teaching Hospitals, Ontario, Canada, 1961*: Section II – Supplement, Toronto: Maternal and Child Health Branch, Department of Health.

Oppenheimer, Jo (1990) "Childbirth in Ontario: The Transition from Home to Hospital in the Early Twentieth Century", in Katherine Arnup, André Levesque and Ruth Roach Pierson (eds), *Delivering Motherhood: Maternal Ideologies and Practices in the Nineteenth and Twentieth Centuries*, New York and London: Routledge: 51–74.

Osborne, Thomas (1994) "On Anti-Medicine and Clinical Reason", in Colin Jones and Roy Porter (eds), *Reassessing Foucault: Power, Medicine and the Body*, New York: Routledge: 28–47.

Osborne, Thomas (1991) "The Doctor's View: Clinical and Governmental Rationalities in Twentieth Century General Medical Practices", unpublished Ph.D. thesis: Brunel University, UK.

Österreichisches Biographisches Lexikon 1815–1950. Vienna (1983) Bd. VIII, "Pfaundler von Hadermur, Meinhard": 27.

Palmer, A.C. (1928) *Child Life Investigations: The Cause of Foetal Death in 144 Cases*, Medical Research Council, Special Report Series, No. 188, London: HMSO.

Parton, Nigel (1998) "Risk, Advanced Liberalism and Child Welfare: The Need to Rediscover Uncertainty and Ambiguity", *British Journal or Social Work*, 28: 5–27.

Parton, Nigel (1996) "Social Work, Risk and 'the Blaming System'", in Nigel Parton (ed.), *Social Theory, Social Change and Social Work*, London and New York: Routledge: 4–18.

Pearce, Neil (1996) "Traditional Epidemiology, Modern Epidemiology and Public Health", *American Journal of Public Health*, 86, 5: 678–683.

Peller, Sigismund (1979) *Not in My Time: The Story of a Doctor*, New York: Philosophical Library.

Peller, Sigismund (1967) *Quantitative Research in Human Biology and Medicine*, Bristol: John Wright and Sons.

Peller, Sigismund (1948) "Mortality, Past and Future", *Population Studies*, 1, 4: 405–456.

Peller, Sigismund (1944) "Studies on Mortality since the Renaissance", Part 2, *Bulletin of the History of Medicine*, 16: 362–381.

Peller, Sigismund (1943) "Studies on Mortality since the Renaissance", Part 1, *Bulletin of the History of Medicine*, 13: 427–461.

Peller, Sigismund (1936) *Der Gebürtstod*, Leipzig and Vienna: Deuticke.

Peller, Sigismund (1931a) "Abortus und Geburtenrückgang", *Der Praktische Arzt*, 18: 514–515.

Peller, Sigismund (1931b) "Erfordert die Bekämpfung der prä-, intra-, und postnatalen Sterblichkeit: medizinische oder sozialmedizinische Reformen", *Zentralblatt für Gynäkologie*, 5: 268–275.

Peller, Sigismund (1931c) "Das intrauterine Wachstum und soziale Einflüsse", *Zeitschrift für Konstitionslehre*, 10: 307–320.

Peller, Sigismund (1931d) "Fehlgeburt und Bevölkerungsfrage", *Biologische Heilkunst*, 39: 639–640.

Peller, Sigismund (1930) "Zum Problem der Tötgeburten und der Neugeborenensterblichkeit", *Klinische Wochenschrift*, 9, 18 (3 May): 844–848.

Peller, Sigismund (1924) "Zur Kenntnis der autonomen Wachstumskraft des Fötus", *Deutsche Medizinische Wochenschrift*, 18: 582–583.

Peller, Sigismund (1923a) "Die Säulingssterblichkeit nach dem Kriege", *Wiener Klinische Wochenschrift*, 45: 799–801.

Peller, Sigismund (1923b) "Die Säulingssterblichkeit nach dem Kriege", *Wiener Klinische Wochenschrift*, 47: 836–837.

Peller, Sigismund and Friedrich Bass (1924) "Die Rolle exogener Faktoren in der intrauterinen Entwicklung des Menschen mit besonderer Berücksichtigung der Kriegs- und Nachkriegverhältnisse", *Archiv für Gynäkologie*, 122: 208–238.

Perkins, Barbara Bridgman (1994) "Intensity, Stratification and Risk in Prenatal Care: Redefining Problems and Restructuring Levels", *Women and Health*, 21, 1: 17–31.

Perkins, Barbara Bridgman (1993) "Rethinking Perinatal Policy: History and Evaluation of Minimum Volume and Level-of-Care Standards", *Journal of Public Health Policy*, 14, 3: 299–319.

Petchesky, Rosalind (1987) "Foetal Images: The Power of Visual Culture in the Politics of Reproduction", in Michelle Stanworth (ed.), *Reproductive Technologies: Gender, Motherhood and Medicine*, Cambridge: Polity Press: 57–80.

Petersen, Alan R. (1996) "Risk and the Regulated Self: The Discourse of Health Promotion as Politics of Uncertainty", *Australian and New Zealand Journal of S ociology*, 32, 1: 44–57.

Petersen, Alan and Deborah Lupton (1996) *The New Public Health*, London: Sage.

Pfaundler, Meinhard von (1936) "Studien über Frühtod, Geschlechtsverhältnis und Selektion", *Zeitschrift für Kinderheilkunde*, 57: 185–227.

Pfaundler, Meinhard von (1924) "Physiologie, Ernährung und Pflege des Neugeborenen, einschließlich des Lebensschwachen", in A. Döderlein (ed.), *Handbuch der Geburtshilfe*, Munich: J.F. Bergmann, Vol. 2: 563–860.

Pfaundler, Meinhard von and Arthur Schlossmann (eds) (1908) *The Diseases of Children: A Work for the Practicing Physician*, Philadelphia: J.B. Lippincott.

Pollock, Leonard J. (1974) "Injuries to Unborn Children", *Alberta Law Review*, 12: 266–274.

Polzer, Jessica (2005) "Choice as Responsibility: Genetic Testing as Citizenship through Familial Obligation and the Management of Risk", in Robin Bunton and Alan Petersen (eds), *Genetic Governance: Health, Risk and Ethics in the Biotech Era*, London: Taylor & Francis: 79–92.

Porter, Dorothy (ed.) (1997) *Social Medicine and Medical Sociology in the Twentieth Century*, Amsterdam and Atlanta, GA: Rodopi.

Porter, Dorothy and Roy Porter (1988) "What Was Social Medicine?", *Journal of Historical Sociology*, 1, 1: 90–106.

Potter, Edith (1952) *Pathology of the Fetus and the Newborn*, Chicago: The Year Book Publishers.

Potter, Edith L. (1947) *Rh: Its Relation to Congenital Hemolytic Disease and to Intragroup Transfusion Reactions*, Chicago: The Year Book Publishers.

Potter, Edith (1943) "The Importance of the Post Mortem Examination of the Fetus and Newly Born Infant", *American Journal of Clinical Pathology*, 13, 3: 133–138.

Potter, Edith (1940) "Fetal and Neonatal Deaths", *Journal of the American Medical Association*, 115, 12: 996–1000.

Potter, Edith (1938) "Postmortem Examination of Stillborn and of Newly Born Infants", *Archives of Pathology*, 25, 5: 607–636.

Potter, Edith and Fred Adair (1943) "Clinical-Pathological Study of the Infant and Fetal Mortality for a Ten-Year Period at the Chicago Lying-In Hospital", *American Journal of Obstetrics and Gynecology*, 45, 6: 1054–1065.

Potter, Edith and Fred Adair (1940) *Fetal and Neonatal Death*, Chicago: University of Chicago Press.

Potter, Edith L. and Fred Adair (1939) "Factors Associated with Fetal and Neonatal Deaths", *Journal of the American Medical Association*, 112, 16: 1549–1556.

Potter, Edith L. and William Jack (1953) "The Responsibility of the Obstetrician in Perinatal Mortality", *The Surgical Clinics of North America*, Chicago Number, Philadelphia and London: W.B. Saunders.

Priest, George (1990) "The New Legal Structure of Risk Control", *Daedalus*, 119: 207–227.

Proceedings of the World Population Conference, Rome, 31 August–10 September 1954 (1955) Summary Report. New York: United Nations, Department of Economic and Social Affairs.

Quéniart, Anne (1988) "De l'insecurité à la solicitude: La médicalisation de la grossesse et ses effets sur le vécu des femmes", in Peta Tancred-Sheriff (ed.), *Feminist Research: Prospect and Retrospect*, Kingston and Montreal: McGill-Queens: 24–35.

Rabinow, Paul (2002) *French DNA: Trouble in Purgatory*, Chicago: University of Chicago Press.

Rabinow, Paul (1996a) "Artificiality and Enlightenment: From Sociobiology to Biosociality", in Paul Rabinow, *Essays in the Anthropology of Reason*, Princeton, NJ: Princeton University Press: 91–112.

Rabinow, Paul (1996b) *Making PCR: A Story of Biotechnology*, Chicago and London: University of Chicago Press.

Rabinow, Paul and Talia Dan-Cohen (2005) *A Machine to Make a Future: Biotech Chronicles*, Princeton, NJ and Oxford: Princeton University Press.

Rabinow, Paul and Nikolas Rose (2003) "Thoughts on the Concept of Biopower Today", http://www.molsci.org/files/Rose_Rabinow_Biopower_Today.pdf (accessed October 2005).

Rapp, Rayna (1999) *Testing Women, Testing the Fetus: The Social Impact of Amniocentesis in America*, New York: Routledge.

Rapp, Rayna (1995) "Risky Business: Genetic Counselling in a Shifting World", in Jane Schneider and Rayna Rapp (eds), *Articulating Hidden Histories*, Berkeley: University of California Press: 175–189.

Rapp, Rayna (1989) "The Power of 'Positive' Diagnosis: Medical and Maternal Discourses on Amniocentesis", in Karen L. Michaelson and contributors (eds), *Childbirth in America: Anthropological Perspectives*, South Hadley, MA: Bergin and Garvey: 103–16.

Report of the Task Force on the Implementation of Midwifery in Ontario (1987) Toronto: Ontario Ministry of Health and Ontario Ministry of Colleges and Universities. Appendix 1: "A History of Midwifery in Canada": 197–227.

Rhéaume, Jean (1990) "Daigle: un oubli des questions de droit civil et constitutionnel", *Revue Générale de droit*, 21, March: 151–167.

Rhoden, Nancy K. (1986) "The Judge in the Delivery Room: The Emergence of Court-Ordered Cesarians", *California Law Review*, 74: 1951–2030.

Richardson, Janice (2004) "Feminist Perspectives on the Law of Tort and the Technology of Risk", *Economy and Society*, 33, 1: 98–120.

Richardson, Ruth (1987) *Death, Dissection and the Destitute*, London and New York: Routledge and Kegan Paul.

Rivet, Michèle (1978) "La situation juridique de l'enfant non encore né au Canada (Droit civil)", in *Travaux du treizième Colloque international de droit comparé* (held at Ottawa (Canada) 3 and 4 October 1975), Ottawa: Éditions de l'Université d'Ottawa: 72–93,

Roberts, Barbara (1988) *Whence They Came: Deportation from Canada, 1900–1935*, Ottawa: University of Ottawa Press.

Roberts, Dorothy E. (1997) *Killing the Black Body*, New York: Vintage Books.

Roberts, Dorothy E. (1991) "Punishing Drug Addicts Who Have Babies: Women of Color, Equality, and the Right of Privacy", *Harvard Law Review*, 104, 7: 1419–1482.

Robertson, Ann (2001) "Biotechnology, Political Rationality and Discourses on Health Risk", *Health*, 5, 3: 293–309

Robertson, Ann (2000) "Risk, Biotechnology and Political Rationality: Lessons from Women's Accounts of Breast Cancer Risks", in Fiona Miller, Lorna Weir, Roxanne Mykitiuk, Patricia Lee and Sari Tudiver (eds), *The Gender of Genetic Futures: The Canadian Biotechnology Strategy, Women and Health*, Toronto: National Network on Environments and Women's Health (Health Canada Centre of Excellence in Women's Health). Working Paper Series, York University: 63–75.

Rose, Nikolas (2001) "The Politics of Life Itself", *Theory, Culture and Society*, 18, 6: 1–30.

Rose, Nikolas (1999) *Powers of Freedom: Reframing Political Thought*, Cambridge: Cambridge University Press.

Rose, Nikolas (1996) *Inventing Our Selves*, New York: Cambridge University Press.

Rose, Nikolas (1994a) "Expertise and the Government of Conduct", *Studies in Law, Politics and Society*, 14: 359–397.

Rose, Nikolas (1994b) "Medicine, History and the Present", in Colin Jones and Roy Porter (eds), *Reassessing Foucault: Power, Medicine and the Body*, New York: Routledge: 48–72.

Rose, Nikolas (1985) *The Psychological Complex: Psychology, Politics and Society in England, 1869–1940*, London: Routledge and Kegan Paul.

Rose, Nikolas and Peter Miller (1992) "Political Power beyond the State: Problematics of Government", *British Journal of Sociology*, 43, 2: 173–205.

Rose, Nikolas and Carlos Novas (2004) "Biological Citizenship", in Aihwa Ong and Stephen J. Collier (eds), *Global Assemblages: Technology, Politics, and Ethics as Anthropological Problems*, Malden, MA: Blackwell: 439–463.

Rose, Nikolas and Mariana Valverde (1998) "Governed by Law?", *Social and Legal Studies*, 7, 4: 541–551.

Rosen, George (1958) *The History of Public Health*, New York: MD Publications.

"Round Table Discussion on Prematurity" (1936) Proceedings of the First Annual Meeting of the American Academy of Pediatrics, *Journal of Pediatrics*, 8, 1: 104–121

Routh, Amand (1913–1914) "Discussion on the Need for Research in Ante-Natal Pathology", *Proceedings of the Royal Society of Medicine* 7, Part 2, Obstetrical and Gynaecological Section: 263–277.

Royal Commission on Aboriginal Peoples (1996) *Report of the Royal Commission on Aboriginal Peoples*, Ottawa: Minster of Supply and Services Canada. Available at http://www.ainc-inac.gc.ca/rcap/sgmm_e.html (accessed April 2006).

Rutman, Deborah (2000) "Misconceiving Ms. G.: A Critical Discourse Analysis of the Canadian Print Media's Coverage of the Case of Ms. G.", in Deborah Rutman, Marilyn Callahan, Audrey Lundquist, Suzanne Jackson and Barbara Field, *Substance Use and Pregnancy: Conceiving Women in the Policy-Making Process*, Ottawa: Status of Women Canada: 47–72.

Rutman, Deborah, Marilyn Callahan, Audrey Lundquist, Suzanne Jackson and Barbara Field (2000) *Substance Use and Pregnancy: Conceiving Women in the Policy-Making Process*, Ottawa: Status of Women Canada.

Rutman, Deborah, Suzanne Jackson and Barbara Field (2000) "Policy Impacts: Part I – Perspectives of Substance-Using Women, Aboriginal Women and Human Service Practitioners; Part II – Aboriginal Women Speak Out", in Deborah Rutman, Marilyn Callahan, Audrey Lundquist, Suzanne Jackson and Barbara Field, *Substance Use and Pregnancy*, Ottawa: Status of Women Canada: 73–138.

Rutman, Leonard and Andrew Jones (1981) *In the Children's Aid: J.J. Kelso and Child Welfare in Ontario*, Toronto: University of Toronto Press.

Sackett, David L., Sharon E. Strauss, W. Scott Richardson, William Rosenberg and R. Brian Haynes (2000) *Evidence-Based Medicine: How to Practise and Teach EBM*, Edinburgh: Churchill Livingstone.

Sandall, Jane, Ivy Lynn Bourgeault, Wouter J. Meijer and Beate A. Schüecking (2001) "Deciding Who Cares: Winners and Losers in the Late Twentieth Century", in Raymond DeVries, Cecilia Benoit, Edwin van Reijlingen and Sirpa Wrede (eds), *Birth by Design: Pregnancy, Maternity Care and Midwifery in North America and Europe*, New York and London: Routledge: 17–138.

Santello, Deborah M. (1988) "Maternal Tort Liability for Prenatal Injuries", *Suffolk University Law Review*, XXII, 747–777.

Sawicki, Jana (1991) *Disciplining Foucault*, New York: Routledge.

Saxell, Lee (1994) "Perceptions of Risk: Women's and Midwives Experience of Nulliparous Pregnancy over 35", unpublished M.A. thesis, Thames Valley University, UK.

Schissel, Bernard and Terry Wotherspoon (2003) *The Legacy of School for Aboriginal People*, Don Mills, Ontario: Oxford University Press.

Schmidt, Glen (1997) "The Gove Report and First Nations Child Welfare", in Jane Pulkingham and Gordon Ternowetsky (eds), *Child and Family Policies: Struggles, Strategies and Options*, Halifax, Nova Scotia: Fernwood.

Schroedel, Jean Reith and Paul Peretz (1994) "A Gender Analysis of Policy Formation: The Case of Fetal Abuse", *Journal of Health Politics, Policy and Law*, 19, 2: 335–360.

Schücking, Beate and Clarissa Schwarz (2002) "Technisierung der 'normalen' Geburt: Interventionen im Kreißsaal", in U. Paravicini and C. Reidel (eds), *Dokumentation Forschungsprojekte 1. bis 3. Förderrunde 1997–2001, Wissenschaftliche Reihe NFFG*, Vol. 1, Hannover: 237–255.

Schulte, F.J., R. Michaelis and R. Nolte (1967) "Meinhard von Pfaundler and the History of Small-for-Dates Infants", *Developmental Medicine and Child Neurology*, 9: 511.

Serbin, William B. (1928) "A Report on 320 Fetal Postmortems at the Chicago Lying-In Hospital", *American Journal of Obstetrics and Gynecology*, 15: 682–685.

Shaffer, Martha (1994) "Foetal Rights and the Regulation of Abortion", *McGill Law Journal*, 39, March: 58–100.

Shaw, Margery W. (1984) "Conditional Prospective Rights of the Fetus", *The Journal of Legal Medicine*, 5, 1: 63–116.

Shroff, Farah M. (ed.) (1997) *The New Midwifery: Reflections on Renaissance and Regulation*, Toronto: Women's Press.

Simon, Jonathan (1993) *Poor Discipline: Parole and the Social Control of the Underclass, 1890–1990*, Chicago and London: University of Chicago Press.

Simon, Jonathan (1988) "The Ideological Effects of Actuarial Practices", *Law and Society Review*, 22, 4: 772–800.

Simon, Jonathan (1987) "The Emergence of a Risk Society: Insurance, Law, and the State", *Socialist Review*, 95: 61–89.

Simpson, John W. and Geppert, Leo J. (1951) "The Responsibility of the Obstetrician to the Fetus", *American Journal of Obstetrics and Gynecology*, 62, 2: 1062–1070.

Singer, Brian and Lorna Weir (forthcoming, 2007) "Politics and Sovereign Power: Considerations on Foucault", *European Journal of Sociology*, 10, 2.

Singh, Navsharan (1997) "Of Victim Women and Surplus Peoples: Reproductive Technologies and the Representation of 'Third World' Women", *Studies in Political Economy*, 52: 155–173.

Silverton, Louise (1993) *The Art and Science of Midwifery*, Hemel Hempstead, Hertfordshire: Prentice Hall.

"Sir Dugald Baird's Birthday Present" (1980) *Lancet*, (Oct. 18): 839–840.

Skocpol, Theda (1992) *Protecting Soldiers and Mothers: The Political Origins of Social Policy in the United States*, Cambridge, MA: Harvard University Press.

Slovic, Paul (1987) "Perceptions of Risk", *Science*, 236: 280–285.

Smith, Clement A. (1946) *The Physiology of the Newborn Infant*, Springfield, IL: Charles C. Thomas.

Smith, Dorothy (1990) *The Conceptual Practices of Power*, Toronto: University of Toronto Press.

Smith, Jeremiah (1917) "Surviving Fictions", *Yale Law Journal*, XXVII, 2: 147–166.

Society of Obstetricians and Gynecologists of Canada (1997) "Midwifery", Clinical Practice Guidelines, Policy Statement No. 66.

Society of Obstetricians and Gynecologists of Canada (1995) "Healthy Beginnings: Guidelines for Care during Pregnancy and Childbirth", unpublished.

Stabile, Carol A. (1992) "Shooting the Mother: Fetal Photography and the Politics of Disappearance", *Camera Obscura*, 28: 119–205.

Stouffer, Martha (1994) "Foetal Rights and the Regulation of Abortion", *McGill Law Journal*, 39, 1: 58–100.

Strathern, Marilyn (1992) *After Nature: English Kinship in the Late Twentieth Century*, Cambridge: Cambridge University Press.

Strickland, Ruth Ann and Marcia Lynn Whicker (1995) "Fetal Endangerment Versus Fetal Welfare: Discretion of Prosecutors in Determining Criminal Liability", in Faye Ginsberg and Anna Lowenhaupt Tsing (eds), *Uncertain Terms: Negotiating Gender in American Culture*, Boston, MA: Beacon Press: 55–84.

Susser, Mervyn (1987) "Evolution in the United States after World War II: The Evolution of Technique", in *Epidemiology, Health, and Society: Selected Papers*, New York and Oxford: Oxford University Press: 22–49.

Susser, Mervyn and Ezra Susser (1996) "Choosing a Future for Epidemiology: I. Eras and Paradigms", *American Journal of Public Health*, 86, 5: 668–673.

Sutherland, Ian (1949) *Stillbirths: Their Epidemiology and Social Significance*, London: Oxford University Press.

Swift, Karen J. (1995) *Manufacturing Bad Mothers: A Critical Perspective on Child Neglect*, Toronto: University of Toronto Press.

Tait, Caroline (2000) *A Study of the Service Needs of Pregnant Addicted Women in Manitoba*, Winnipeg: Prairie Women's Health Centre for Excellence, http://www.gov.mb.ca (accessed October 2005).

Task Force on the Implementation of Midwifery in Ontario (1987) *Report of the Task Force on the Implementation of Midwifery in Ontario. Toronto: The Task Force*, Appendix I, "A History of Midwifery in Canada": 197–227.

Tateishi, S.A. (1989) "Apprehending the Fetus *en Ventre Sa Mère*: A Study in Judicial Sleight of Hand", *Saskatchewan Law Review*, 53: 113–141.

Taylor, Janelle (2000) "Of Sonograms and Baby Prams: Prenatal Diagnosis, Pregnancy, and Consumption", *Feminist Studies*, 26, 2: 391–418.

Taylor, Janelle (1992) "The Public Fetus and the Family Car", *Public Culture*, 4, 2: 67–80.

Temkin, Owsei (1956) "Introduction", Soranus' *Gynecology*, Baltimore, MD: Johns Hopkins: xxiii–xlix.

Terris, Milton (1987) "Epidemiology and the Public Health Movement", *Journal of Public Health Policy*, 8, 3: 315–330.

Terris, Milton (1985) "The Changing Relationships of Epidemiology and Society: The Robert Cruikshank Lecture", *Journal of Public Health Policy*, 6, March: 15–36.

Tew, M. (1990) *Safer Childbirth? A Critical History of Maternity Care*, London: Chapman and Hall.

Thompson, Joyce E., Linda V. Walsh and Irwin R. Merkatz (1990) "The History of Prenatal Care: Cultural, Social, and Medical Contexts", in Irwin R. Merkatz and Joyce E. Thompson (eds), *New Perspectives on Prenatal Care*, New York: Elsevier: 9–30.

Thomson, Christine J. (1926) "The Causation of Still-birth and Neo-natal Death", *Journal of Obstetrics and Gynaecology of the British Empire*, XXXIII: 390–418.

Tsing, Anna Lowerhaupt (1990) "Monster Stories: Women Charged with Perinatal Endangerment", in Faye Ginsberg and Anna Lowerhaupt Tsing (eds), *Uncertain Terms: Negotiating Gender in American Culture*, Boston, MA: Beacon Press: 282–299.

Tyson, Holliday, Anne Nixon, Arlene Vandersloot and Kate Hughes (1995) "The Re-emergence and Professionalization of Midwifery in Ontario, Canada", in Tricia Murphy-Black (ed.), *Issues in Midwifery*, Edinburgh: Churchill Livingstone: 163–175.

United Kingdom, Law Commission (1974) *Report on Injuries to Unborn Children*, Law Com. No. 60 Cmnd 5709, London: HMSO.

United Kingdom, Law Commission (1973) *Injuries to Unborn Children*, Published Working Paper No. 47, London: HMSO.

Valaoras, Vasilios (1954) "Foetus, Peri-Natal and Infant Mortality", *Proceedings of the World Population Conference, 1954b*, Rome, 31 August–10 September, Papers Vol. 1: 323–335.

Valverde, Mariana (1991) *The Age of Light, Soap and Water: Moral Reform in English Canada, 1885–1925*, Toronto: McClelland and Stewart.

Vickers, Brian (1988) *In Defence of Rhetoric*, Oxford: Clarendon.

Wadhera, Surinder and Jill Strachan (1993) *Selected Infant Mortality and Related Statistics, Canada, 1921–1990*, Canadian Centre for Health Information, Statistics Canada, C93-099362–4E.

Van Wagner, Vicki (2004) “Why Legislation?: Using Regulation to Strengthen Midwifery”, in Ivy Lynn Bourgeault, Cecilia Benoit and Robbie Davis-Floyd (eds), *Reconceiving Midwifery*, Montreal: McGill-Queens: 71–90.

Van Wagner, Vicki (1988) “Women Organizing for Midwifery”, *Resources for Feminist Research*, 7, 13: 115–118.

Wallace, Helen M., Edwin M. Gold, Margaret A. Losty and Herbert Rich (1952) “Criteria for the Evaluation of the Quality of Maternity and New-Born Care”, *American Journal of Public Health*, 42, 5: 499–507.

Ward, W. Peter (1981) “Unwed Motherhood in Nineteenth-Century English Canada”, *Historical Papers*, Canadian Historical Association: 34–56.

Warren, Kenneth R. and Laurie Foudin (2001) “Alcohol-Related Birth Defects – The Past, Present, and Future”, *Alcohol Research and Health*, 25, 3: 153–158.

Warry, Wayne (1991) “Ontario’s First People: Native Children”, in Laura Johnson and Richard Barnhorst (eds), *Children, Families and Public Policy in the 90s*, Toronto: Thompson: 207–230.

Weiler, Karen M. and Katherine M. Catton (1976) “The Unborn Child in Canadian Law”, *Osgoode Hall Law Journal*, 14, 3: 643–659.

Weindling, Paul (1989) *Health, Race and German Politics between National Unification and Nazism, 1870–1945*, Cambridge: Cambridge University Press.

Weindling, Paul (1988) “Fascism and Population in Comparative European Perspective”, *Population and Development Review*, 14 (Supplement): 101–121.

Weir, Lorna (1998a) “Cultural Intertexts and Scientific Rationality: The Case of Pregnancy Ultrasound”, *Economy and Society*, 27, 2 and 3: 249–259.

Weir, Lorna (1998b) “Pregnancy Ultrasound and Maternal Discourse”, in Margrit Shildrick and Janet Price (eds), *Vital Signs: Feminist Reconfigurations of the Bio/Logical Body*, Edinburgh: Edinburgh University Press: 78–101.

Weir, Lorna (1997) “Negotiating Risk: Ontario Midwives and the Ontario Antenatal Record”, in Mariana Valverde (ed.), *New Forms of Governance: Theory, Practice, Research*, Toronto: Centre of Criminology: University of Toronto: 16–18.

Weir, Lorna (1996) “Recent Developments in the Government of Pregnancy”, *Economy and Society*, 25, 3: 372–392.

Wertz, Richard W. and Dorothy Wertz (1989) *Lying-In: A History of Childbirth in America*, New Haven, CN: Yale University Press.

Williams, J.W. (1915) “The Limitations and Possibilities of Prenatal Care”, *Journal of the American Medical Association*, 64: 95–101 .

Williams Obstetrics (1956) Nicholson J. Eastman, New York: Appleton-Century-Crofts. 11th edn.

Williams Obstetrics (1950) Nicholson J. Eastman, New York: Appleton-Century-Crofts.10th edn.

Willson, Prentiss (1924) “The Problems of Prenatal, Natal, and Neonatal Mortality”, *International Clinics*, 1, 34th Series: 41–47.

Winfield, P.H. (1942) “The Unborn Child”, *Cambridge Law Journal*, 8: 76–91.

Winfield, P.H. (1926) “The History of Negligence in the Law of Torts”, *Law Quarterly Review*, 42: 184–201.

Wiskott A. (1966) "In Memoriam: Meinhard von Pfaundler", *Zeitschrift für Kinderheilkunde*, 96: 97–105

Women's Legal, Education and Action Fund (1996) *Equality and the Charter – Ten Years of Feminist Advocacy*, Toronto: Emond Montgomery Publications.

World Health Organisation, Regional Office for Europe (1985) *Having a Baby in Europe: Report on a Study*, Copenhagen.

World Health Organisation (1967) *Manual of the International Statistical Classification of Diseases, Injuries, and Causes of Death*, Vol. 1. Geneva: World Health Organisation.

World Health Organisation (1957) *Manual of the International Statistical Classification of Diseases, Injuries, and Causes of Death*, Vol. 1. Geneva: World Health Organisation.

Wright, Peter W.G. (1988) "Babyhood: The Social Construction of Infant Care as a Medical Problem in England in the Years around 1900", in Margaret Lock and Deborah Gordon (eds), *Biomedicine Examined*, Dordrecht: Kluwer: 299–330.

Wynne, Brian (1996) "May the Sheep Safely Graze? A Reflexive View of the Expert–Lay Knowledge Divide", in Scott Lash, Bronislow Szerszynski and Brian Wynne (eds), *Risk, Environment and Modernity*, London: Sage: 44–83.

Wynne, Brian (1995) "Public Understanding of Science", in S. Jasanoff, G.E. Markle, J.C. Petersen and T. Pinch (eds), *Handbook of Science and Technology Studies*, London: Sage: 361–388.

Yankauer, Alfred (1957) "The Public Health Aspects of Perinatal Mortality", New York *State Medical Journal*, 57, 3: 2499–2502.

Yankauer, Alfred (1953) "An Evaluation of Prenatal Care and its Relationship to Social Class and Social Disorganization", *American Journal of Public Health*, 43, 8: 1001–1010.

Yankauer, Alfred (1950) "The Relationship of Fetal and Infant Mortality to Residential Segregation: An Inquiry into Social Epidemiology", *American Sociological Review*, 15, 5: 644–648.

Yerushalmy, J. (1938) "Neonatal Mortality by Order of Birth and Age of Parents", *American Journal of Hygiene*, 28: 244–270.

Index